THE GREAT TREE OF LIFE

THE GREAT TREE OF LIFE

DOUGLAS E. SOLTIS
Florida Museum of Natural History, University of Florida,
Gainesville, FL, United States

PAMELA S. SOLTIS
Florida Museum of Natural History, University of Florida,
Gainesville, FL, United States

ACADEMIC PRESS
An imprint of Elsevier

Academic Press is an imprint of Elsevier
125 London Wall, London EC2Y 5AS, United Kingdom
525 B Street, Suite 1650, San Diego, CA 92101, United States
50 Hampshire Street, 5th Floor, Cambridge, MA 02139, United States
The Boulevard, Langford Lane, Kidlington, Oxford OX5 1GB, United Kingdom

Notices
Knowledge and best practice in this field are constantly changing. As new research and
experience broaden our understanding, changes in research methods, professional practices, or
medical treatment may become necessary.

Practitioners and researchers must always rely on their own experience and knowledge in
evaluating and using any information, methods, compounds, or experiments described herein.
In using such information or methods they should be mindful of their own safety and the safety
of others, including parties for whom they have a professional responsibility.

To the fullest extent of the law, neither the Publisher nor the authors, contributors, or editors,
assume any liability for any injury and/or damage to persons or property as a matter of products
liability, negligence or otherwise, or from any use or operation of any methods, products,
instructions, or ideas contained in the material herein.

British Library Cataloguing-in-Publication Data
A catalogue record for this book is available from the British Library

Library of Congress Cataloging-in-Publication Data
A catalog record for this book is available from the Library of Congress

ISBN: 978-0-12-812553-3

For Information on all Academic Press publications
visit our website at https://www.elsevier.com/books-and-journals

Working together
to grow libraries in
developing countries

www.elsevier.com • www.bookaid.org

Publisher: Andre Gerhard Wolff
Acquisition Editor: Anna Valutkevich
Editorial Project Manager: Michelle Kubilis
Production Project Manager: Sreejith Viswanathan
Cover Designer: Mark Rogers

Typeset by MPS Limited, Chennai, India

DEDICATION

To Katie and Sarah

CONTENTS

ACKNOWLEDGMENTS

Many people have helped with this book. Katie and Sarah Soltis read the chapters and provided input. Sarah also helped research topics and located important figures; her insights into art were also very important for Chapter 1, Tree of Life in Ancient Human Culture and Art. Special thanks to Evgeny Mavrodiev for researching selected topics, finding much-needed references, and assisting with some of the figures. Elena Mavrodieva drew one of the figures for Chapter 2, History of the Modern Concept of the Tree of Life. Many people provided valuable discussions and ideas. David Blackburn and Akito Kawahara provided information on cryptic species. Jack Payne provided the inspiration regarding the work of Kozlovsky. Ken Sassaman provided helpful discussions on Mayan and other cultures and their use of Tree of Life imagery. Sangtae Kim provided images of *Amborella*. Jeremy Lichstein helped with ecological references, figures, and ideas relating to ecological cascades. Brett Scheffers helped with information and a figure on the huge footprint of global climate change. Heather Rose Kates provided a figure as well as input on crop improvement. Robert Thacker provided information on sponges. David Steadman was invaluable in his help with information on bird extinction and conservation and provided an image used here. David Reed provided input and a tree figure on lice. Julie Allen and Kurt Neubig provided images for the study of biodiversity of Florida. Ryan Folk was instrumental in research and images for Saxifragales. James Rosindell provided high-resolution images of OneZoom. Stephen Smith read part of the text for Chapter 3, Building the Tree of Life: A Biodiversity Moonshot, and also provided Tree of Life images used here. Ross Purnell helped inspire the text on sport fishing and conservation in Chapter 5. Naziha Mestaoui, James Rosindell, Matt Gitzendanner, and Rob Guralnick have been amazing collaborators on the Tree of Life projection, *One Tree, One Planet*. Similarly, thanks to James Oliverio, Tim Difato, and others at the University of Florida's Digital Worlds Institute for showing the importance of animation in teaching the Tree of Life. Our research has been funded for nearly four decades by the National Science Foundation, and we truly appreciate this support. Thanks also to the University of Florida for funding. Many others have also been of great help, including our students, postdocs, and lab visitors. Finally, a special thanks to the many colleagues with whom we have worked over many years in building and using the Tree of Life. That journey inspired this book.

INTRODUCTION

Look deep into nature, and then you will understand everything better.
Albert Einstein 1951

We all seem to have a fascination with our family trees—who were my ancestors? How am I related to others? There is a deep understanding across all of us—that knowing these relationships matters. Certainly, one part of that importance is realizing "Who am I?"—and where do I fit in as part of the grand scheme of human history of relationships? "I have native American ancestry, or German ancestors, or an African ancestor (fill in the blank)—I did not know that."

But today more than ever, as we have a deeper appreciation of the genetic underpinning that defines us all, we also realize that a family tree has a predictive value! That is, we know that if a certain trait, a high occurrence of a disease, is part of my own family history, we have the concern that we may have that trait as well. More specifically, if a close relative had a certain disease that can be inherited, say a certain cancer, we understand that there is a good probability that I may have inherited the genes for that trait.

This same logic or reasoning applies to the biggest family tree of them all, the great Tree of Life that unites all species on our planet. That is, all organisms on Earth are part of a family tree, too—species also have close relatives, whether they be whales, fish, butterflies, lice, or flowering plants. We can similarly ask—How are the millions of species on the planet Earth related? Where do we humans fit on this tree of all life? And because relationships matter, including at the level of the tree of all life, we can also ask, what are the many ways that this knowledge of relationships may benefit humans?

The concept of the Tree of Life has a rich history. As nicely summarized by Lima (2014, p. 16), "the concept of a Tree of Life ... has been one of the most widespread and long-lasting archetypes of our species." Most ancient cultures have a Tree of Life concept in which all organisms on Earth are connected, intertwined, as are branches and leaves on a tree. In this context, the Tree of Life had rich spiritual and religious meaning, linking the underworld, life, and the heavens. The Tree of Life is depicted frequently in ancient art.

The ancient Greeks had a firm concept of organismal relationships—the beginnings of a more modern Tree of Life view. The Greek philosopher Aristotle suggested that all species form a great hierarchy that he envisioned from simple to the complex, in a linear fashion.

The modern concept of a Tree of Life that we follow today in which all species are related via common descent and part of a great tree of all life, dates to the time of Darwin. In fact, the only illustration in Darwin's famous book, *On the Origin of Species*, is not a finch or a map of the Galapagos islands . . . but a simple tree of relationships.

But surprisingly, until just a few years ago, there had never been a single, comprehensive tree of all life—one that included all of the roughly two million named species of life on Earth. Despite decades of research and numerous studies on various subgroups of organisms—birds, mammals, flowering plants, bacteria, etc.—we still lacked a comprehensive tree of all life. Why? Quite simple, building large trees of relationship is not easy—it is extremely difficult, rivaling some of the major challenges in physics in terms of mathematical complexity.

In fact, 30 years ago when our lab and others started building "big trees" of just a few hundred species, we were told this was an impossible task—computationally impossible. Hence, building the Tree of Life was in many ways "a moonshot" for biodiversity science—a grand challenge that many deemed impossible to meet. But through a perfect storm of technology, computer power, and the DNA sequencing revolution, as well as team work, building the tree of all named life became feasible. DNA data allowed scientists to readily obtain data for thousands of species for use in building these big trees. Building big trees of relationships is computationally extremely challenging. Major developments in computer power as well as new algorithms were major contributors to success. These breakthroughs in several areas at the same time made it possible for researchers to build the first rough draft Tree of Life for all of the 2.3 million named species on planet Earth!

But it is also important to stress that this first tree of all named life is really just a starting point. The goal now is to improve the tree—our knowledge of life on our own planet is surprisingly poor. DNA data are an essential tool for establishing relationships among species and building large trees. But, we only have DNA data for perhaps 17% of the species included in the first efforts to build a comprehensive tree. Hence, most relationships remain poorly understood.

Of equal or greater urgency in realizing an improved Tree of Life is that the number of species named and included in the first draft tree represents only a small fraction of the number of species thought to inhabit our planet. In fact, scientists estimate that there may be 10 million more species of eukaryotic life on Earth that are still undiscovered, unnamed! And many, many millions more undiscovered prokaryotic species (bacteria and archaea). In other words, by far most species of life on Earth are not included in this initial Tree of Life. The bottom line—we really do not know that much about life on Earth!

There are numerous practical examples of the utility of the Tree of Life. Again—relationships matter. Trees of relationships are the first line of defense in combatting diseases and in drug discovery. The Tree of Life is also important for crop improvement. Knowledge of wild relatives of our cultivated crops may be a crucial source of genes for drought tolerance and immunity to disease. The Tree of Life is important for conservation. Understanding how species are related is essential for assessing which rare lineages are highly distinct and may therefore merit extra conservation concern. Trees of relationship are now essential tools in conservation and ecology. In a rapidly changing world, ecologists can use the Tree of Life to predict how species may respond to higher temperatures or to less water … Why? Because relationships matter … closely related species will respond in a similar way to changes in the environment. The biodiversity represented in the Tree of Life is essential for clean air and water and a livable planet.

At the same time, we are in a biodiversity crisis. We are watching a planet die—with species disappearing at an alarming rate—approximately 1000 times the normal background rate and rising. And disturbingly, many species are lost before they are even discovered and named. How many of these disappearing species have (or may have held) the key to medical and other economic benefits to our own species, not to mention the value of these species in their own right as valuable components of ecosystems, providing diverse benefits such as clean air, clean water, as well as enhanced mental health (take a walk in the woods to discover the latter benefit). Never has the time been so critical—to act now to appreciate and embrace the importance of the Tree of Life, of biodiversity.

Our goal in writing this book is straightforward—call attention to the biggest family tree of them all—the great Tree of Life of all species on Earth. We want to stress that appreciation of the Tree of Life is not just biological, but spiritual and that appreciation appears to be as old as our own species. From a practical standpoint, we will stress why knowledge of that tree is important for human well-being and survival. Relationships matter! Biodiversity matters! The Tree of Life is an essential tool with many benefits for our own species. The downstream practical benefits of knowledge of the Tree of Life for humans are numerous—including the discovery of medicines, curing diseases, improving our crops, and even predicting the global response of species to a rapidly changing climate.

For these reasons, and others, we will also argue that it is crucial that we work to improve our understanding of how organisms on Earth are related to each other. Imagine a world with a more rapid discovery of medicines, curing of disease, improvement of crops, and prediction of how organisms will respond to a rapidly changing environment.

With the threat of extinction facing more species than ever before in human history, the time for rapid action is now. It is imperative to improve our understanding of the Tree of Life for human health and well-being. But the task is daunting—the current version of the Tree of Life is a poor understanding of organismal relationships. Only a fraction of species is now represented by DNA data—most species are simply placed where we think they belong, based on descriptions of their appearance. Furthermore, although new species continue to be described at the rate of $\sim 14{,}000$/year—which seems substantial—at this rate of description it will take at least another 900 years to describe all species on Earth. Do we really want to wait that long? It will take a committed effort that includes scientists and the public to significantly improve our understanding of life sooner than later—it takes a village to build and improve the Tree of Life.

We hope that this book will impact not only the scientific community, but also, more importantly, the public and policy makers. Collectively, through a major biodiversity initiative, we can and must greatly improve our knowledge of the Tree of Life over the next few decades. If we do nothing, the best-case scenario is that we will lose important medicines, foods, and perhaps ecosystem benefits (clean water). The worst-case scenario, with the rapid increase of extinction on our planet and the challenges faced, in particular, in developing countries, we risk unprecedented damage to our ecosystems and loss of biodiversity—including many thousands of unnamed species of a potential value to our species.

CHAPTER 1

Tree of Life in Ancient Human Culture and Art

Behind the Man is the Tree of Life...

A. E. Waite 1911

INTRODUCTION

The Hopi from the southwestern United States have elaborate stories about a past world in which its inhabitants exploited nature, to an extent that led to their own destruction (Mohawk, 2001). In many cultures, it was a crime to cut down a tree due to its spiritual and medicinal importance. Trees have been a fundamentally important symbol, with diverse meanings throughout human history. Lima (2014) has argued that trees and their structure have provided an organizing principle in many aspects of human life, ranging from systems of law to areas of science even beyond biology. Trees represent an entity of tremendous importance to humans around the world and act as a metaphor for the connectivity of all life (including our own species)—the Tree of Life, a powerful symbol, with diverse, yet interconnected, connotations.

This book focuses on the Tree of Life in several ways, beginning with its use as a human symbol throughout history to represent the connectivity of all life, with the phrase Tree of Life representing that all-encompassing connectivity. Significantly, throughout most of human history, our species, *Homo sapiens*, has felt a deep connection to the other species on our planet—this connection was largely what we might term spiritual and is evident in ancient religions and art, with depictions of trees used to represent that intertwining of all life. Most human cultures used tree symbols and had a Tree of Life concept that was essentially spiritual, mystical, and religious (Liya, 2004; Parpola, 1993; Giovino, 2007). This sense of connectivity to all of life through much of human history is one that often seems absent in the modern world. The Tree of Life concept is explored here not only in the ancient, cultural sense but also from

The Great Tree of Life
DOI: https://doi.org/10.1016/B978-0-12-812553-3.00001-1

a more modern evolutionary perspective—one used since the time of Darwin to indicate the clear genealogical connection of our species to all other species on our planet.

TREE OF LIFE: MULTIPLE MEANINGS

For clarity, the term Tree of Life has been used in multiple ways, and there are also other related terms. A very similar term used by scholars is the term "world tree," which refers to the use by many, highly diverse ancient cultures of an actual image of a large tree as an important symbol or motif. Many human societies had a view of a giant tree that was strongly linked to the concept of a Tree of Life connecting all living things. In many cases, the world tree symbol was also connected with what has been termed the tree of knowledge, in which trees connect heaven to the underworld (Berkurt, 1998; Liya, 2004; Giovino, 2004, 2007). These intertwined or linked trees—tree of knowledge, Tree of Life, the world tree motif—show the immense importance of the tree metaphor in human societies around the world.

Depictions of trees have appeared commonly in cultures around the world and throughout human history (reviewed in Liya, 2004; Lima, 2014). Liya refers to the tree as "a universal symbol found in the myths and sacred writings of all peoples." The connection of humans to trees had diverse elements that were both spiritual and mystical (Liya, 2004; Parpola, 1993; Giovino, 2004, 2007).

But the connection also reflects the clear practical utility of trees to our species. Trees have been used throughout human history for food, shelter, fire, and as a source of medicines (most medicines trace to plants), as well as mind-altering substances. For example, the acacia tree, a member of the bean family, had a special place in ancient Egyptian spirituality. To the Serer people in Africa, the Somb tree (another bean family member) was revered as the ancestor of all trees on Earth. Both trees also appear to have medicinal use, showing a connection of practical utility and spirituality. To the Navajo people of the southwestern United States, their Tree of Life assumed the shape of maize (corn) (Lima, 2014). Today, we often forget that trees have been a central facet of human development for their many uses. Put simply, human survival and the rise of human societies, civilizations, and religions have depended on trees, and that dependency continues to this day (Anderson, 1967; Baker, 1965; Shery, 1972).

Not only are trees fundamental for human survival, they have many characteristics that were revered by humans. Trees are long-lived organisms, suggesting longevity, and often attain large stature, evoking power and connections to other worlds. Tall trees reaching skyward suggested to early humans a connection to the heavens and a spiritual link to life after death. Multiple cultures felt this way, including native people in the Americas and in northern Europe. The growth of an enormous long-lived tree from a small seed also had a deep spiritual meaning (Liya, 2004). Furthermore, the connection of leaves to twigs to stems to branches and to trunk represented the connection of all life, and a massive trunk with roots that penetrate deep into the Earth suggested a connection to the underworld. All of these factors contribute to making the tree a universal symbol of life and connectivity (Liya, 2004; Lima, 2014).

THE TREE OF LIFE: A METAPHOR FOR BIODIVERSITY AND CONNECTIVITY

The Tree of Life metaphor is one with a long, diverse, and rich history. As summarized beautifully by Morgenstern (2003), "since time immemorial plants have played in human cultures throughout time a key role in human spirituality. Their sublime beauty, entrancing scents … have suggested a connection with 'the other world,' the nonmaterial world of Gods and sprites, demons and devils." Cultures throughout human history have viewed trees as a symbol representing the connection between different levels of existence—the heavens via the crown and the underworld via the roots (Liya, 2004), with the connectivity of leaves and branches representing the connectivity of all life. The psychologist Carl J. Jung (not to be confused with Carl G. Jung, the founder of analytical psychology) was intrigued by the frequency with which humans of all cultures and backgrounds had dreams of trees. His long-term research of the subject prompted him to conclude that the tree is a fundamental symbol, a part of the human unconscious (Jung, 1967; Liya, 2004).

People throughout history have had a Tree of Life story or myth deeply imbedded in their culture and religion (Liya, 2004), and many such myths have striking similarities across very different people living in very distant parts of the world. Sometimes that myth involved a mystical tree or tree image, and in others there was a religious or ceremonial connection to actual tree species. Some Tree of Life concepts held by ancient cultures maintained that eating the fruit of the Tree of Life would result in eternal life (e.g., the Christian Bible story; see below). In cultures in both Asia and North America, the Tree of Life was considered to be present and growing at the center of the world. This all-important Tree of Life was protected by the supernatural. In these cultures, our own species was believed to be descended from this Tree of Life. Furthermore, destruction of the Tree of Life—i.e., cutting down the Tree of Life—would result in the end of our species, more specifically, the end of human reproduction and ultimately the end of *Homo sapiens* (e.g., Liya, 2004). As noted, the Hopi talk about a past world in which the inhabitants over-exploited nature, leading to their own destruction (Mohawk, 2001). These are important cultural myths to reflect on today, given the many threats to and destruction of the "Tree of Life"; human-mediated extinction of species making up the Tree of Life will ultimately have dire consequences for our own species (see Wilson, 2016; see Chapter 6).

Many cultures, including the Maya, native North Americans, and those of northern Europe, had a religious belief centered on animism, and many indigenous people today, including certain Amazonian people, still follow this practice. That is, all things, including living organisms and various physical entities such as rocks, rivers, buildings, and artifacts, are considered living. In this sense, these people are spiritually connected to all other organisms ... spiritually part of the Tree of Life and the entities around them. They place a tremendous importance on social connections between the nonhuman and human domains (e.g., Bird-Davis, 1999; Brown and Emery 2008; Hornborg, 2006; Viveiros de Castro, 2004). In these cultures, diverse species, including plants as well as nonliving forms such as rocks, may be approached and considered as objects with which one can communicate—not simply inert objects.

This is not to say that ancient societies necessarily lived in perfect harmony with nature, that they were conservationists or ecologically minded ... but they certainly felt a spiritual connection to other life forms. As discussed in Chapter 6, the human capacity to drive organisms to extinction has a long history, one that may also be as old as *Homo sapiens* (Kolbert, 2006, 2014; Cafaro, 2015). "The evidence suggests that human beings have been disastrous for other species ever since we evolved into something recognizably ourselves." (Cafaro, 2015, p. 255). The modern human-dominated period in which we are now living, the Anthropocene (see Chapter 6), "has been the age of death for many other species and 'the killing' has been going on for thousands of years" (Cafaro, 2015, p. 255).

There is clear evidence that many native people did not live in ecological balance with the world they inhabited (see Krech, 2000). Focusing on native people of the Americas, Krech, in *The Ecological Indian*, takes a critical look at claims that native people had strong ecological tendencies and sensible sustainability practices. He maintains that the image held by modern societies that native people in the Americas lived in harmony with nature is misleading. Native Americans used fire to alter habitats, yielding, for example, a vast mosaic of open fields and forest in portions of eastern North America. Krech's view has received support, as well as considerable criticism, and stimulated debate (Harkin and Lewis, 2007). For example, one such criticism suggests that Krech does not give enough attention to the difference between traditional (pre-European contact) versus post-traditional native cultures (Mohawk, 2001), the argument being that many native people (pre-Western contact) viewed nature very differently than we do today.

ANCIENT VIEWS OF THE TREE OF LIFE

Artistic depictions of the Tree of Life are often found throughout ancient Assyria. Stylistic depictions of trees are commonly found throughout ancient Mesopotamia, occurring as far back as 6000 BP (years Before Present). These stylistic trees of life often consisted of a trunk and a crown surrounded by a network of lines and were considered by scholars to have clear cultural and religious significance: trees connected all life and further connected our species to the heavens and to the underworld (Parpola, 1993; Giovino, 2007). Furthermore, by 4000 BP, these artistic depictions of what is considered to represent the Tree of Life are found more widely geographically, including in ancient Egypt, Greece, and the Indus Valley (Parpola, 1993), but the tree motif during this timeframe in these areas varied in depiction. Simple forms often consisted only of a trunk and a crown surrounded by a network of lines. But during this time period, more elaborate versions of the Tree of Life appeared in which the tree was adorned with depictions of animals, humans, and even figures of the supernatural (Fig. 1.1). Contemporaneously, the Tree of Life was also associated with royalty and had become an imperial symbol. Scholars indicate that it is not completely clear what the tree symbol represented as the meaning of the trees is not discussed in ancient texts (Giovino, 2007). Some suggest it represents the "Tree of Life" as told in the Bible in Genesis (see below), known as the kiskanu from cuneiform

Figure 1.1 Assyrian Tree of Life; photograph from artifact in the British Museum. Protective spirits with eagle heads with the Sacred Tree of Life, 865—860 BC. *From Wikipedia Free Commons.*

writings. Another interpretation is that the Assyrian tree symbolizes a real tree, a depiction of a date palm, which was an important tree for the many products it provided. But these interpretations need not be mutually exclusive—human societies often attached spiritual meaning to actual tree species while also embracing a symbolic Tree of Life. Researchers suggest that despite modern uncertainties regarding the actual meaning of tree symbols to Assyrians, the widespread use of the Tree of Life motif indicated that the symbolic nature of these tree depictions must have been commonly known—having both "mystical and religious symbolism"; to Assyrians, they perhaps represented in part "the divine world order" (reviewed in Parpola, 1993).

In the world of ancient Egypt (starting at roughly 5000 BP until 2300 BP), the Tree of Life was also a central theme, a metaphysical concept that symbolized the hierarchical series of events that resulted in the formation of all life. In ancient Egypt, the Tree of Life was represented as a series of spheres that, to the ancient people of Egypt, represented not only the process of creation but also the method and orderly manner of events by which this process occurred. Interestingly, the placement of our own species is not a main focus of these creation myths; instead, the focus is the establishment of cosmic order (Pinch, 2004). As in many ancient cultures, actual tree species were associated with Tree of Life mythology. The acacia tree, a member of the bean family (*Acacia nilotica*; now

Vachellia nilotica), had a special place in ancient Egyptian religion—it was considered the Tree of Life.

Approximately 2600 BP, in the area of ancient Iran, a religion founded by the prophet Zoroaster was practiced. Trees were powerful symbols in the religion referred to as Zoroastrianism; the Tree of Life represented the center of the zoster philosophy. Two trees, Mashya and Mashyana, were considered the ancestors of all living things. According to Zoroastrianism, the Tree of Life was considered the way to heaven, and Tree of Life thinking was deeply spiritual for its followers (Liya, 2004). Trees represented the connectivity of all life and hence were revered by those following Zoroastrianism. To kill a tree was considered a sin.

Continuing with the area of the Middle East, the Tree of Life plays an important role in both Judaism and Christianity. Most Jewish and Christian people know well the Bible story in Genesis that begins with Adam and Eve in the Garden of Eden where two trees are present. One of the trees is the Tree of Life. But much better known is the second tree in the Garden of Eden—the Tree of Knowledge of Good and Evil. In Genesis, both the Tree of Life and the Tree of Knowledge of Good and Evil were considered to be planted together. In Genesis 2:9: "And out of the ground made the Lord God to grow every tree that is pleasant to the sight, and good for food; the Tree of Life also in the midst of the garden, and Tree of knowledge of Good and Evil." Well known to Jews and Christians is the passage from Genesis in which Adam and Eve consume the fruit of the Tree of Knowledge and are then expelled by God from the Garden of Eden—to be removed from proximity to the Tree of Life to prevent Adam and Eve from eating the fruit of the Tree of Life and then "live forever" (Liya, 2004); from Genesis 3:23−25, "So he drove out the man; and he placed at the east of the garden of Eden Cherubims, and a flaming sword which turned every way, to keep the way of the Tree of Life."

There has been debate as to whether the Tree of Knowledge of Good and Evil and the Tree of Life are actually the same tree (reviewed in Mettinger, 2007). For example, the Quran tells a similar story to that given in Genesis, but the Quran refers to only a single tree (*Qurán* 20:118; Liya, 2004). But Bible scholars indicate that the two trees are almost always dealt with separately and considered not related to each other. However, attention is typically given by writers to the Tree of Knowledge of Good and Evil, whereas in contrast, the Tree of Life is hardly noted (reviewed in Mettinger, 2007). But the Tree of Life plays a prominent role and is mentioned elsewhere in the Bible. In fact, the Tree of Life is noted 11 times in the Bible: in Proverbs (3:18; 11:30; 13:12; 15:4), 2 Esdras (2:12; 8:52), 4 Maccabees (18:16), and Revelation (2:7; 22:2, 14, and 19). Significantly, the Bible also ends with the Tree of Life (Liya, 2004). In Revelation, when Christ returns to Earth, it is stated that the Tree of Life will be growing beside the water of life, "bearing twelve crops of fruit, yielding its fruit every month—and the leaves of the tree are for the healing of the nations" (Revelation 22:2). Significantly, there-fore, the Bible begins and ends with the Tree of Life, again reflecting the significant role it must have played in spirituality of cultures.

Tree of Life concepts are also found in other parts of Africa beyond Egypt. The Serer people of what is now Senegal in western Africa trace their origins back to at least the 11th century. In Serer culture, trees and the Tree of Life were, and remain, central to their religion. In the Serer religion, trees were the first things created on Earth, and the Tree of Life was a central religious concept. To the Serer people, as to other ancient religions, trees were sacred and were actually given religious status. In the Serer creation myth, as in the creation myths of many other cultures across the globe, actual trees played a central role. The Somb tree (*Prosopis africana*, or African mesquite, a member of the bean family) and the Saas tree (*Faidherbia albida*, also a member of the bean family) were both viewed as trees of life (Gravrand, 1990). The Somb tree was not only considered the first tree on Earth but was also thought to have given rise to all other plants (Gravrand, 1990; Niangoran-Bouah, 1987). The Somb tree was also a symbol of immortality.

In Asia, the Tree of Life played a central role to Turkic people, the diverse people who occupied a geographically large area in parts of Asia as well as eastern Europe. Turkic people trace to approximately the sixth century AD. They first inhabited parts of Central Asia to Siberia and were highly nomadic (www.britannica.com/topic/Turkic-peoples). The Turkic religion is considered to be rooted in Zoroastrianism (see above) (Beckwith, 2009; Roxburgh, 2005; Findley, 2005); so, not surprisingly, the Tree of Life was a key symbol and the focal point of Turkic mythol-ogy. Turkic people known as the Altai maintained that humans are actu-ally the descendants of trees. According to another group of Turkic people, the Yakuts, the goddess White Mother, who created Earth (Findley, 2005), is present at the base of the Tree of Life; in Yakut mythology, the Tree of Life reached into the heavens where the branches were occupied by various species of life.

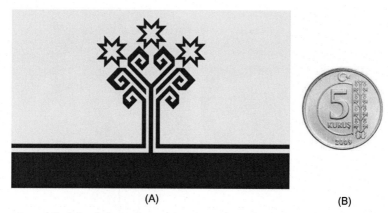

(A)   (B)

Figure 1.2 (A) Flag of Chuvash Republic showing stylized Tree of Life; the Chuvash represent a Turkic ethnic group. The Tree of Life has a long history of importance to Turkic people; the Chuvash flag shows that this importance as a symbol continues today. *From On the state symbols of the Chuvash Republic: Chǎvash en, 24 July-02 Aug. (No. 30-31), 1−7.* (B) A stylized representation of the Tree of Life on the 5 Kurus coin of Turkey. *From Wikipedia Free Commons.*

The Tree of Life continues as a symbol to Turkic people today. It is a common motif in Turkish carpets. Most striking is the presence of the Tree of Life in the flag of the Chuvash Republic (Fig. 1.2; https://syr-maepon.livejournal.com/60690.html), an autonomous republic within the Russian Federation home to the Chuvash people, a Turkic ethnic group. A stylized representation of the Tree of Life is also present on the 5 Kurus coin of Turkey (Fig. 1.2b), which is similar in its tree depiction to those seen in ancient Egypt and Assyria (see Fig. 1.1).

The concept of an enormous tree at the center of the universe surrounded by a number of worlds was common in cultures found across northern Eurasia in the Middle Ages and was central to their mythology (Davidson and Ellis, 1993). Sacred trees played an important role in areas occupied by Germanic people who shared a belief, noted previously for other early cultures, that humans originated from trees (Simek, 2008). The best-known Tree of Life myth from northern Europe traces to the ancient Norse people, who referred to the Tree of Life as Yggdrasil (Fig. 1.3). Evidence for the myth of Yggdrasil appears as early as the 13th century, where it is mentioned in the *Poetic Edda* (Dronke, 1997; Larrington, 1999). According to these compilations, Yggdrasil was an immense tree. As with many other cultures, the Norse people associated their Tree of Life with an actual tree species, typically thought to have been an ash, or by others, a yew. This tree was considered sacred and the

Figure 1.3 Yggdrasil. This is the best-known Tree of Life myth from across northern Europe; it traces to the ancient Norse (Scandinavian) people. *From Wikipedia Free Commons.*

center of the universe, with its branches thought to extend into the heavens. Yggdrasil was also surrounded by extensive lore with diverse creatures living within it (Lindow, 2001; Dronke, 1997). Odin, the key god in Norse mythology, has sometimes been associated with the Norse Tree of Life and even with the name Yggdrasil itself, but scholars indicate that the precise meaning of the name Yggdrasil remains unclear (Simek, 2008). According to Norse mythology, nine worlds existed around Yggdrasil (Davidson and Ellis, 1993).

Davidson and Ellis (1993) note parallels between the Norse Tree of Life lore and similar mythology across areas of northern Eurasia. While the concept of Yggdrasil is ancient, the importance of sacred trees remained a feature of northern European people for centuries. As recently as the 19th century, sacred trees (referred to as warden trees) were revered and venerated in areas of Germany and Scandinavia; they were thought to bring good luck, and offerings were sometimes made to these trees (Davidson and Ellis, 1993).

The Tree of Life was also a focal point of mythology in the Americas. The Tree of Life and tree imagery were a centerpiece of mythology among the Iroquois and apparently other native people of North America, as reviewed in Parker (1912). The Iroquois Confederation or League comprised five nations (Mohawk, Onondaga, Cayuga, Seneca, and Oneida; the Tuscarora were later added as a sixth nation). The league may have formed between 1100 and 1450 (estimates vary) and dominated much of northeastern North America prior to European contact (Mann and Fields, 1997; Johansen, 1995). In historical accounts of the Iroquois, there are references to the "tree of peace"—the Iroquois referred to actions involving peace using the metaphor of a tree (Parker, 1912). Parker gives many examples of Iroquois reference to the peace tree.

In the mythology of all Iroquois nations, there is also reference to the "tree of the upper world" (celestial tree), although details differ among tribes (Fig. 1.4). Parker noted that the fullest description of this upper world tree is from a translation from a Seneca tribal member to a missionary, and this account remains striking today:

> There was a vast expanse of water. Above it was the great blue arch of air ...
> In the clear sky was an unseen floating island sufficiently firm to allow trees to
> grow upon it, and there were men-beings there. There was one great chief who
> gave the law to all the...beings on the island. In the center of the island there
> grew a tree so tall that no one of the beings who lived there could see its top.
> On its branches flowers and fruit hung all the year round. The beings who lived
> on the island used to come to the tree and eat the fruit and smell the sweet
> perfume of the flowers.

Figure 1.4 Celestial Tree of Life of the Iroquois from eastern North America (fig. 64. page 617). Iroquois tree myths and symbols. *From Parker, A.C., 1912. Certain Iroquois tree myths and symbols. Am. Anthropol. 14, 608–620.*

The Iroquois myth of the Tree of Life continues with reference to a skymother who fell from the upper world and landed on the wings of birds who placed her on the back of a turtle (details in Parker, 1912). In this and similar myths from other tribes in eastern North America, the Tree of Life then grew from the back of the turtle. According to the Seneca, the Tree of Life is described as the "tree whose branches pierce the sky and whose roots extend to the waters of the underworld" (Parker, 1912). Another tribe from eastern North America, the Delaware, stated about the Tree of Life, that "among whose branches men had grown" (Parker, 1912), a clear symbolic connection of humans to the Tree of Life. Depictions of symbolic trees also appeared commonly in the art and clothing of the Iroquois (Fig. 1.5).

To the ancient Maya, the Tree of Life represented the connectivity among three worlds—heaven, earth, and the underworld (a world tree). To the Maya, the Tree of Life was represented physically by the kapok

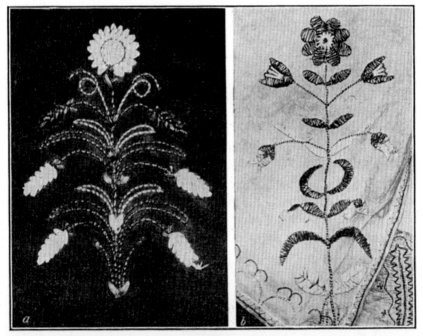

Figure 1.5 Seneca porcupine quill embroidery Tree of Life (fig. 67b). Iroquois tree myths and symbols. *From Parker, A.C., 1912. Certain Iroquois tree myths and symbols. Am. Anthropol. 14, 608–620.*

tree (*Ceiba pentandra*), a member of the cotton family (Malvaceae), and one of the largest and tallest trees in their world. Tree of Life depictions are found in Mayan architecture to the late Formative (Preclassic) Period (Formative Period = 1800 BC to AD 200) (Miller and Taube, 1993) (Fig. 1.6). Other cultures from Mesoamerica, for example, the Aztec, also had similar depictions of the Tree of Life as a world tree (e.g., Lima, 2014).

Figure 1.6 Mayan, Mesoamerican sacred tree, drawing based on the largest Stela (monument) at the Izapa ruins, Tapachula, Mexico. *From Wikipedia Free Commons.*

Research suggests that modern descendants of the ancient Maya, as well as people in the ancient Mayan civilization, negotiated with "the animate forest" (Brown and Emery, 2008). This follows on the view (above) that many native people were "animistic"—entities including rocks, plants, and rivers were essentially alive, entities with which one communicated. Humans and forests were both animate (alive), and ceremonial activities alleviated the potential danger when one agent entered the domain of the other. Hunting, for example, was an encroachment on the animate forest. Brown and Emery (2008) review evidence for hunting shrines, places used in ceremonial fashion to negotiate with the animal guardian associated with the forest landscape.

FROM TREES TO LADDERS: GREEKS AND THE LADDER OF LIFE

The ancient Greeks had a firm knowledge of biodiversity and a concept of the Tree of Life. Through the work of a number of well-known Greek scholars and their students, for the first time, we see the organization of nature into discrete categories—or at least the first time it is recorded in writing. This work originating in ancient Greece really represents the beginning of humankind constructing biological classification. In many ways, what the Greeks started culminated in the classification system of Carolus Linnaeus, a system still largely used today.

Rather than envisioning a tree-like series of connections, the ancient Greeks proposed that all of life could be placed in some type of linear hierarchy among species. For example, Plato (423−347 BC) considered the problem of "natural kinds," that is, how entities could be organized, and he recognized four classes (Kraut, 2015). His student Aristotle (384−322 BC) introduced the fundamental concepts or categories of genus (*genos*) and species (*eidos*)—but these were very broad categories compared to our usage of these terms today (Barnes, 1984). Plato was an "essentialist"; that is, he believed that each entity had an essential feature, an essence, that defines the type of entity in question: a group is real if it has an essence and a final cause unique to itself. In this sense, a species is an expression of its own individual essence: "it is in its essence that a thing possesses being or existence." As part of this essentialist view, species never change—they are immutable entities. This view of species remained central to human thinking for hundreds of years, until the work of Charles Darwin (see Chapter 2).

Aristotle also recognized major categories of animals—vertebrates and invertebrates—further subdividing each into categories that are still recognized today (e.g., birds, mammals, fish in the former) (Barnes, 1984; reviewed in Shields, 2016a,b; Bergstrom and Dugatkin, 2011, Rhodes and Trevor, 1974). A student of Aristotle, Theophrastus (c. 371—c. 287 BC), followed the tradition of classification proposed by Aristotle and organized all plants into four great categories—trees, shrubs, subshrubs, and herbs. This was the first systematic classification of plants and was followed through the Middle Ages. For this and other contributions, Theophrastus has been referred to as the Father of Botany—and a precursor to Linneaus (Ierodiakonou, 2016).

Plato and then Aristotle also envisioned that all life was part of a hierarchy, with the simplest forms at the bottom and more complex at the apex. This was Aristotle's *Scala Naturae* (reviewed in Mayr, 1982; Lovejoy, 1936, 1964), in which the tree or hierarchy was considered analogous to the rungs of a ladder. In this view, entities were arranged in a linear series, with our species occupying the apex—with all other life leading towards our own species, the culmination or perfection of the hierarchy.

This view of life as a ladder culminating in humans would be further developed by Porphyry of Tyre, a philosopher in the Roman Empire (AD 234—305). He largely followed the thinking of Aristotle in terms of the essence or "concept of substance" that can be attributed to diverse entities including life. Mirroring Aristotle, two of his components were a broadly defined genus and species (Strange, 1992; Gracia and Newton, 2016). Porphyry ultimately developed and presented a classification based on the original work of Aristotle. Porphyry does not appear to have developed this concept into an actual tree-like depiction (Franklin, 1986), but ultimately his classification was transformed into a tree by medieval scholars who followed his writings (Fig. 1.7). Nonetheless, these trees were referred to as Porphyrian trees. The Tree of Porphyry became widely used and well known (Gracia and Newton, 2016).

This view of a ladder of nature is one that would be long held by Western civilization and is very much at odds with the views of other cultures that our species is just a part of an overall Tree of Life. It also differs from the modern scientific view of *Homo sapiens* as only one tip on the Tree of Life. Nonetheless, this view of life with humans at the top would long impact Western thought on the organization of biodiversity. As reviewed by Lovejoy (1936, 1964), this view would ultimately give rise to a religious interpretation of the "Great Chain of Being" or

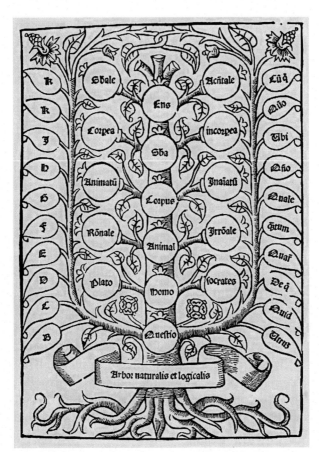

Figure 1.7 Porphyrian Tree, depicting life as a ladder culminating in humans. *From Wikipedia Free Commons.*

Scala Naturae (published in 1745 in the *Traité d'insectologie* by Charles Bonnet), which maintains a strict hierarchy from god to angels to human (kings and nobles occupying places above commoners) to animals, plants, and even rocks and minerals. In this view, humans occupied the top of the hierarchy of life on Earth. In many ways, this view still permeates human thought today—that we are not part of the Tree of Life, but somehow separate, at the apex of a ladder of perfection.

Trees remain impressive, awe-inspiring organisms to most humans to this day. In the United States, we have National Parks and National Monuments dedicated largely to tree species because of their magnificent features, whether it be size, age, or both (Redwood and Sequoia National Parks for size and age; the Ancient Bristlecone Pine Forest for age). In China, ginkgos and the dawn redwood, *Metasequoia* (a relative of both *Sequoia* (redwood) and *Sequoiadendron* (giant redwood)) are similarly revered and protected for their age and size. Humans continue to find refuge in forests for spiritual renewal—it just feels good to be surrounded by trees (see Chapter 5). If intelligent life forms from other worlds were to come to Earth, perhaps the organisms on our planet that would most fascinate and intrigue them would be trees, not our own species.

REFERENCES

Anderson, E., 1967. Plants, Man, and Life. University of California Press, Berkeley, CA.

Baker, H.G., 1965. Plants and Civilization. Wadsworth Publishing Company, Belmont, CA.

Barnes, J. (Ed.), 1984. Complete Works of Aristotle, The Revised Oxford Translation. Princeton University Press, Princeton, NJ.

Beckwith, C.I., 2009. Empires of the Silk Road: A History of Central Eurasia From the Bronze Age to the Present. Princeton University Press, Princeton, NJ.

Bergstrom, C.T., Dugatkin, L.A., 2011. Evolution. W. W. Norton & Company, New York.

Berkurt, W., 1998. Creation of the Sacred: Tracks of Biology in Early Religion. Harvard University Press, Boston, MA.

Bird-Davis, N., 1999. "Animism" revisited: personhood, environment, and relational epistemology. Curr. Anthropol 40 (S1), 67−91.

Brown, L.A., Emery, K.F., 2008. Negotiations with the animate forest: hunting shrines in the Guatemalan Highlands. J. Archaeol. Method. Theory 15, 300−337.

Cafaro, P., 2015. Recent Books on species extinction. Biological Conservation 181, 245−257.

Davidson, H., Ellis, R., 1993. The Lost Beliefs of Northern Europe. Routledge, London, UK.

Dronke, Ursula, 1997. The Poetic Edda (Trans.) Mythological Poems, Volume II. Oxford University Press.

Findley, C.V., 2005. The Turks in World History. Oxford University Press.

Franklin, J., 1986. Aristotle on species variation. Philosophy 61, 245−252.

Giovino, M.D., 2007. The Assyrian Sacred Tree: A History of Interpretations. Academic Press Fribourg, Vandenhoeck and Ruprecht, Gottingen, Germany.

Gracia, J., Newton, L., 2016. Medieval theories of the categories. In: Zalta, E.N. (Ed.), The Stanford Encyclopedia of Philosophy Archive. Metaphysics Research Lab, Stanford University, Stanford, CA.

Gravrand, H., 1990. Pangool: le genie religieux sereer. Les Nouvelles Editions Africaines du Senegal, Senegal.

Harkin, M.E., Lewis, D.R. (Eds.), 2007. Native Americans and the Environment Perspectives on the Ecological Indian. University of Nebraska Press, Lincoln, NE.

Hornborg, A., 2006. Animism, fetishism, and objectivism as strategies for knowing (or not knowing) the world. Ethnos 71, 21−32.

Ierodiakonou, K., 2016. Theophrastus. In: Zalta, E.N. (Ed.), The Stanford Encyclopedia of Philosophy Archive. Metaphysics Research Lab, Stanford University, Stanford, CA.

Johansen, B.E., 1995. Dating the Iroquois Confederacy. Akwesasne Notes New Series, New York, pp. 62−63.

Jung, C.J., 1967. The Philosophical Tree II. On the History and Interpretation of the Tree Symbol, Alchemical Studies, Collected Works of C.G. Jung, 13. Princeton University Press, Princeton, NJ, pp. 272—349.

Kolbert, E., 2006. Field Notes From a Catastrophe: Man, Nature, and Climate Change, first ed Bloomsbury, New York.

Kolbert, E., 2014. The Sixth Extinction: An Unnatural History, first ed Henry Holt and Co, New York.

Kraut, R., 2015. Plato. In: Zalta, E.N. (Ed.), The Stanford Encyclopedia of Philosophy Archive. Metaphysics Research Lab, Stanford University, Stanford, CA.

Krech, S., 2000. The Ecological Indian: Myth and History. W. W. Norton and Company, N.Y.

Larrington, C., 1999. The Poetic Edda. Oxford University Press, Oxford, UK and New York.

Lima, M., 2014. The Book of Trees: Visualizing Branches of Knowledge. Princeton Architectural Press, New York.

Lindow, J., 2001. Norse Mythology: A Guide to the Gods, Heroes, Rituals, and Beliefs. Oxford University Press, Oxford, UK.

Liya, S., 2004. The use of trees as symbols in the world religions. Solas 4, 41—58.

Lovejoy, A.O., 1936. The Great Chain of Being: A Study of the History of an Idea—The William James Lectures Delivered at Harvard University. Harvard University Press, Cambridge, MA, p. 1933.

Mann, B.A., Fields, J.L., 1997. A sign in the sky: dating the League of the Haudenosaunee. Am. Indian Cult. Res. J. 21, 105—163.

Mayr, E., 1982. The Growth of Biological Thought: Diversity, Evolution, and Inheritance. The Belknap Press of Harvard University Press, Cambridge, MA and London, UK.

Mettinger, T.N.D., 2007. The Eden Narrative: A Literary and Religio-Historical Study of Genesis 2—3. Eisenbrauns, Winona Lake, IN.

Miller, M., Taube, K., 1993. The Gods and Symbols of Ancient Mexico and the Maya: An Illustrated Dictionary of Mesoamerican Religion. Thames and Hudson Ltd, London, UK.

Mohawk, J.C., 2001. Review of the ecological Indian: myth and history by Shepard Krech III. Nat. Soc. Sci. 11, 183—184.

Morgenstern, K., 2003. Plants as Gateways to the Sacred. Sacred Earth Newsletter, Pleasantville, NY.

Niangoran-Bouah, G.G., 1987. L'univers akan des poids à peser l'or (The Akan world of gold weights). Nouvelles Editions Africaines, Abidjan, Republic of Côte d'Ivoire (Ivory Coast).

Parker, A.C., 1912. Certain Iroquois tree myths and symbols. Am. Anthropol. 14, 608—620.

Parpola, S., 1993. The Assyrian tree of life—tracing the origins of Jewish monotheism and Greek philosophy. J. Near East. Stud. 52, 161—208.

Pinch, G., 2004. Egyptian Mythology: A Guide to the Gods. Goddesses, and Traditions of Ancient Egypt. Oxford University Press.

Rhodes, F., Trevor, H., 1974. Evolution (A Golden science guide). Golden Press, Fayetteville, NC.

Roxburgh, D.J., 2005. Turks: A Journey of a Thousand Years, 600-1600. Royal Academy of Arts, London, UK and New York.

Shery, R.W., 1972. Plants for Man, second ed Prentice-Hall, Englewood Cliffs, NJ.

Shields, C., 2016a. Aristotle. In: Zalta, E.N. (Ed.), The Stanford Encyclopedia of Philosophy Archive. Metaphysics Research Lab, Stanford University, Stanford, CA.

Shields, C. (Ed.), 2016b. Aristotle's De Anima. first ed Clarendon Press and Oxford University Press, Oxford, UK and New York.

Simek, R., 2008. A Dictionary of Northern Mythology. Boydell & Brewer, Limited, Rochester, NY.

Strange, S.K., 1992. Ancient commentators on Aristotle. Bloomsbury Academic, New York.

Viveiros de Castro, E., 2004. Exchanging perspectives: the transformation of objects into subjects in Amerindian ontologies. Common Knowledge 10, 463–485.

Wilson, E.O., 2016. Half-Earth. Liveright Publishing Corporation, New York.

FURTHER READING

Budge, E.A.W., 1969. The Gods of the Egyptians; or, Studies in Egyptian Mythology. Dover Publications, New York.

Carter, V.F., 2005. The Turks in World History. Oxford University Press, New York.

Dowden, K., 2000. European Paganism: The Realities of Cult From Antiquity to the Middle Ages. Routledge, London, UK; New York.

Porphyry, 1975. Porphyry the Phoenician, Translation, Introduction, and Notes by Edward W. Warren. Pontifical Institute of Mediaeval Studies, Toronto, Canada.

Sorabji, R. (Ed.), 1992. Porphyry, on Aristotle Categories. Bristol Classical Press, London-Ithaca, NY.

Shepard Ill, K., 1999. The Ecological Indian: Myth and History. W. W. Norton & Co, New York.

Studtmann, P., 2014. Aristotle's categories. In: Zalta, E.N. (Ed.), The Stanford Encyclopedia of Philosophy Archive. Metaphysics Research Lab, Stanford University, Stanford, CA.

CHAPTER 2

History of the Modern Concept of the Tree of Life

...and so by generation I believe it has been with the great Tree of Life, which fills with its dead and broken branches the crust of the earth, and covers the surface with its ever-branching and beautiful ramifications.

Darwin 1859

TREE OF LIFE—DARWIN AND EVOLUTIONARY CONNECTIONS

The modern concept of a Tree of Life that depicts evolutionary relationships dates to the time of Charles Darwin (1809—82; Fig. 2.1), who recognized that organismal evolution meant that all species were linked by descent—so the history of thinking in terms of trees of genealogical connections has a short but rich history. Significantly, there is only one figure in Darwin's (1859) famous book, *On the Origin of Species*. That figure is not a drawing of a finch or a map of the Galapagos Islands but a simple phylogenetic tree (a tree of relationships) drawn by Darwin that depicts relationships among a hypothetical group of organisms (Fig. 2.2A). Darwin famously states, "The affinities of all the beings of the same class have sometimes been represented by a great tree. I believe this simile largely speaks the truth. As buds give rise by growth to fresh buds, and these, if vigorous, branch out and overtop on all sides many a feebler branch, so by generation I believe it has been with the great Tree of Life, which fills with its dead and broken branches the crust of the earth, and covers the surface with its ever-branching and beautiful ramifications." Thus, it is evident from this single figure and this quotation from *On the Origin of Species* that Darwin was also considering fossils (extinct species) in this phylogenetic tree and not just relationships among living species—his view of a great tree linking all species was a Tree of Life and death.

However, Darwin was clearly thinking about phylogenetic trees long before the publication of *On the Origin of Species*. Early sketches in

The Great Tree of Life
DOI: https://doi.org/10.1016/B978-0-12-812553-3.00002-3

Figure 2.1 Photograph of Charles Darwin. *From Wikipedia Free Commons.*

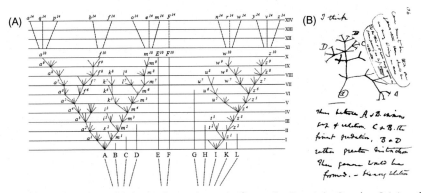

Figure 2.2 (A) Tree of relationships; the only figure in Darwin's *On the Origin of Species. From Darwin, C., 1859. The Origin of Species: By Means of Natural Selection, Or the Preservation of Favoured Races in the Struggle for Life. Cambridge, UK: Cambridge University Press.* (B) Darwin's 1837 tree sketch. *Wikipedia Free Commons.*

Darwin's notebook B from 1837 show what appear to be simple branching patterns that some suggest actually resemble corals, organisms with elaborate branching patterns. These early drawings indicate that Darwin initially may have thought that the coral, whose branching patterns are almost tree-like, was the best metaphor for the branching patterns among living things (http://phylonetworks.blogspot.com/2012/06/charles-dar-wins-unpublished-tree.html). But a later and the most famous sketch from Darwin's notebook B from 1837 clearly shows a branching diagram of relationships illustrating evolutionary connections among organisms (Fig. 2.2B). This famous sketch has even become a popular tattoo! Accompanying this diagram, Darwin writes as a first note "I think," which is followed to the right by the statement "Case must be that one generation then should be as many living as now." This is then followed by "To do this and to have many species in same genus (as is) requires extinction" (http://phylonetworks.blogspot.com/2012/06/charles-dar-wins-unpublished-tree.html). These notes importantly show that Darwin was thinking about a Tree of Life in an evolutionary context. Not only was he thinking about speciation and the relationships of species through time as forming a tree-like branching pattern, but he was also considering the role of extinction in the Tree of Life.

It is important to stress that other biologists from the same time period were also thinking along the same lines that all species of life were con-nected by descent from a common ancestor. In fact, Darwin's grandfather, Erasmus Darwin (1731–1802), was clearly thinking in terms of evolution and a Tree of Life well before Charles. In his book *Zoonomia* (1794–96), Erasmus Darwin stated:

> *Would it be too bold to imagine, that in the great length of time, since the earth began to exist, perhaps millions of ages before the commencement of the history of mankind, would it be too bold to imagine, that all warm-blooded animals have arisen from one living filament, which the great First Cause endued with animality, with the power of acquiring new parts, attended with new propensities, directed by irritations, sensations, volitions, and associations; and thus possessing the faculty of continuing to improve by its own inherent activity, and of delivering down those improvements by generation to its poster-ity, world without end!*

There were also trees of relationships published before Darwin's famous work. The earliest published Tree of Life appears to be that of the French botanist Augustin Augier in 1801. Augier was not a well-known scientist, and his work and depiction of a plant Tree of Life were lost to obscurity until rediscovered by Stevens (1983) who gives an excellent overview of Augier and his plant Tree of Life (Fig. 2.3). Stevens stresses that Augier did not think in terms of evolution; however, although he was certainly influenced by the *Scala Naturae* (the "Great Chain of Being,"

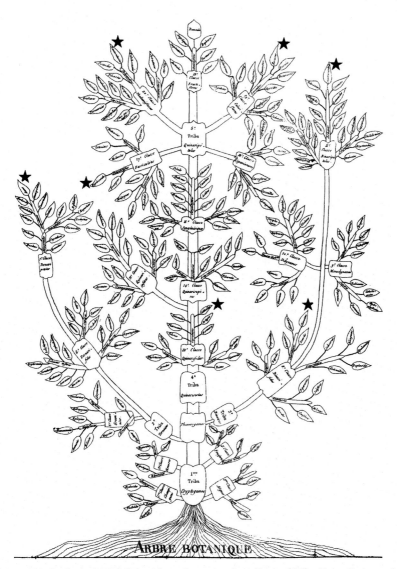

Figure 2.3 Augier's (1801) Arbre Botanique or plant Tree of Life. *From Stevens, P.F., 1983. Augustin Augier's "Arbre Botanique" (1801). A remarkable early botanical representation of the natural system. Taxon 32 (2), 203–211, with permission.*

the hierarchical structure of life, as well as matter; see Chapter 1: Tree of Life in Ancient Human Culture and Art), a dominant view at that time, his depiction of relationships in a tree that more closely resembles a family tree shows the beginning of a movement away from the *Scala Naturae*.

Not long after Augier, the famous scientist Jean-Baptiste Lamarck (1744–1829) published the first known Tree of Life for animals in 1809. Lamarck is best known for his theory of evolution via inheritance of acquired traits, and his tree is also not very tree-like, consisting of separate lines to lineages of animals—that is, the animals were not shown to be linked by common descent.

Charles Darwin's book, *On the Origin of Species* (1859), quickly inspired the famous German biologist Ernst Haeckel (1834–1919; Fig. 2.4A) to depict his view of the Tree of Life in a series of well-known elaborate drawings (1866, 1879–80; Fig. 2.4B). Haeckel's trees are early attempts to summarize the Tree of Life from the scientific perspective and remain inspiring today. Haeckel was a multitalented scientist, and in addition to his detailed depictions of the Tree of Life, he also made beautiful illustrations of organisms that are still appreciated for both their scientific and esthetic qualities. Haeckel was also responsible for the now debunked hypothesis that ontogeny recapitulates phylogeny—that is, the early developmental stages of organisms reveal their evolutionary history.

Darwin's diagram of an evolutionary tree in *On the Origin of Species* and the rich imagery of his text subsequently inspired not only Haeckel, but also several generations of scientists to depict evolutionary relationships using hand-drawn trees. These depictions by scientists who were experts in specific groups of organisms—flowering plants, mammals, or birds—reflected the views of the investigator regarding how those species were related based on years (sometimes a lifetime) of study.

This basic approach of illustrating relationships among organisms continued well into the 1900s with prominent biologists using variants of hand-drawn branching schemes. The well-known American botanist Charles Bessey (1845–1915) depicted a summary of flowering plant relationships in a scheme that resembles a cactus and is still referred to today as Bessey's (1915) "cactus" (Fig. 2.5). Arthur Cronquist (1919–92) and other botanists used a similar approach in the 1970s and 1980s, producing what have been referred to as bubble diagrams (Fig. 2.6).

The diagrams of relationships provided by Haeckel (1866), Bessey (1915), Cronquist (1968, 1981), and others have in common the shared feature that the relationships shown by these authors reflected the "gut feelings" of the scientist. These scientists also focused on how major groups of organisms were related but did not typically depict all species. For example, the flowering plant group Caryophyllales of Cronquist contains thousands of species, but the interrelationships among all of the

(A)

(B)

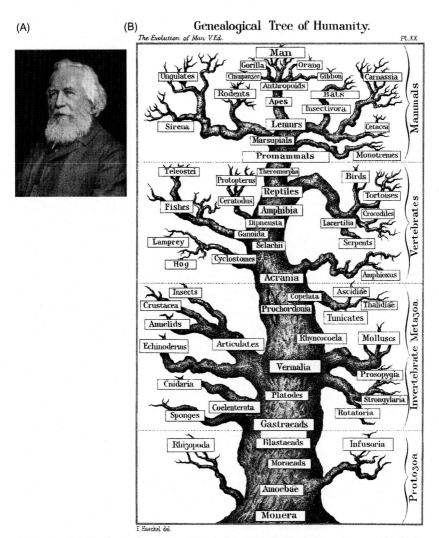

Genealogical Tree of Humanity.

Figure 2.4 (A) Photograph of Ernst Haeckel (1834—1919), famous German biologist. *From Wikipedia Free Commons.* (B) One of Ernst Haeckel's depictions of the Tree of Life. *From The Evolution of Man (1879). Wikipedia Free Commons.*

species are not shown. Furthermore, the "trees" of these authors reflected their interpretation of the morphological characters ("appearance") which they deemed important, but these are not rigorously built trees based on explicit methods.

This approach of subjective tree-building characterized the field of "systematics" (the area of study having the daunting task of naming,

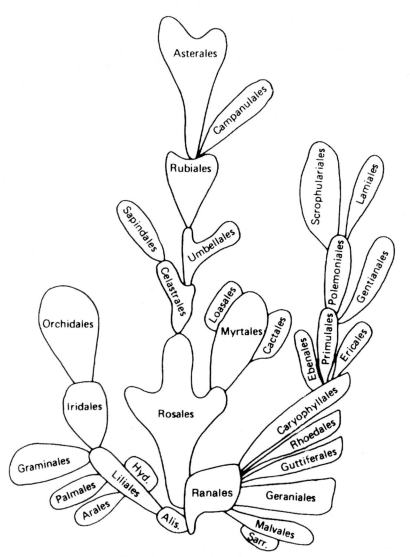

Figure 2.5 Depiction of a summary of flowering plant relationships by American botanist Charles Bessey (1845–1915); the depiction (from Bessey 1915) is often referred to as Bessey's cactus.

describing, and classifying species and determining how all are related) for roughly 100 years from the time of Darwin until the middle of the 1900s. Tree-thinking in this vein was based on expertise but was hardly rigorous from a scientific standpoint or repeatable. It was unclear which features or traits were used to produce an author's view of relationships. There was

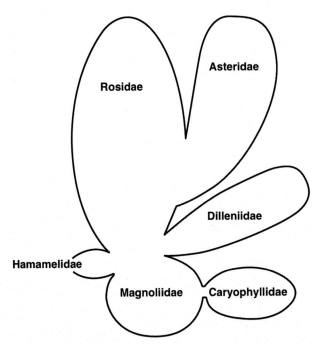

Figure 2.6 Depiction of a summary of flowering plant relationships by American botanist Arthur Cronquist (1919—92). *Redrawn from Cronquist, A., 1981. An Integrated System of Classification of Flowering Plants. New York: Columbia University Press.*

no way for others to test hypotheses of relationships with a data set (none was provided) or repeat the analyses (none were conducted). It was completely subjective.

REVOLUTION: EXPLICIT METHODS FOR CONSTRUCTING TREES OF EVOLUTIONARY RELATIONSHIPS

A tree of relationships is called a *phylogeny.* This will be one of the few big terms used in this book. As reviewed elsewhere in this book, trees of relationships are crucial to modern biology and human existence. Evolutionary trees are so fundamentally important that it is time for the term "phylogeny" to be part of the common, everyday vernacular, just as the terms ecology and ecosystem are widely used and accepted words by nonscientists. Phylogeny is from the Greek word *Phulon,* which literally means "tribe", and "-geny", which refers to the origin—thus, phylogeny refers to the origin of groups.

It is clear from the above discussions that a scientifically rigorous method of building trees of relationships was badly needed—something more than the gut feeling of the investigator, no matter how many years of experience and organisms that investigator might have studied. Surprisingly, although rigorous means of constructing even very simple "phylogenetic trees" were first proposed nearly 70 years ago, such methods have only been commonly used in biology for half that time and only in the past 10 years has it been possible to build really large trees containing thousands of species. This latter breakthrough came about due, in large part, to innovations in technology, including the development of new algorithms, enhanced computer power, and novel DNA sequencing methods (see Chapter 3: Building the Tree of Life: A Biodiversity Moonshot).

It was not until the mid-1900s that well-defined, explicit, repeatable methods and criteria for building phylogenetic trees were developed by scientists. As often happens in science, several people ultimately contributed to this advance independently at about the same time. The most prominent contributions came from Willi Hennig (1913—76) (Fig. 2.7A), a German entomologist. His personal life story is an amazing

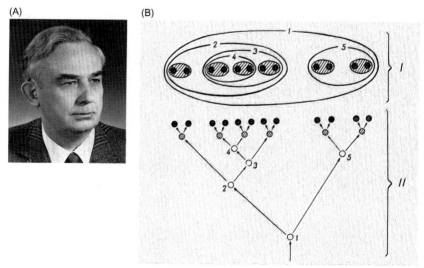

(A)   (B)

Figure 2.7 (A) Photograph of Willi Hennig (1913—76), a German scientist who developed tree-building methods. *From Wikipedia Free Commons.* (B) Simple tree from Hennig. *Redrawn from Fig. 18 of Hennig, W., 1966. Phylogenetic Systematics. Urbana, IL: University of Illinois Press.*

one, a story in its own right. He was a scientist (an entomologist—a person who studies insects) and a German officer in World War II. Professor Hennig is considered the founder of *phylogenetic systematics*, a field with the goal of building trees of relationships among organisms. He outlined his theory of phylogenetic systematics in 1950 in German (Hennig, 1950), and, as a result, it was not widely read initially. It was not until his ideas were published in English (Hennig, 1966) that the impact of his theory of tree building became widely appreciated and ultimately used. Hennig used characters and states of those characters (e.g., the character flower color might have states red, pink, and white) to build trees of relationship (Fig. 2.7B). However, Hennig's ideas were based in part on the earlier ideas of a German botanist, Walter Zimmerman (1892–1980), which Hennig clearly acknowledged (see Donoghue and Kadereit, 1992). An American fern biologist, Warren "Herb" Wagner (1920–2000), independently formulated some of the very same ideas as did Hennig in the mid-1950s, unaware of Hennig's work (reviewed in Wagner, 1980). So, in short, the study of building phylogenetic trees was revolutionized by a number of scientists, Zimmerman, Hennig, and Wagner, all in the mid-1900s, all using general features of morphology (appearance) as characters and character states. It is important therefore to keep in mind that the rigorous building of phylogenetic trees is a relatively new field, without the long history of other sciences such as chemistry or physics.

Hennig more than any other person provided the first explicit mechanisms for constructing trees. He introduced key concepts such as the term *clade*—a group of species composed of an ancestor and all of its descendants (see the simple example in Fig. 2.8). Hennig constructed data sets of characters (e.g., insect wing spots) with each character having different variants or states (e.g., spots present or absent). These data sets of numerous characters and states were the underlying "evidence" to build trees of relationships.

Hennig's novel contribution was that groups of related organisms—clades—could be recognized based on new character states that species had in common which were lacking in related groups—in other words, shared derived states. For example, if one considers wing venation in a group of insects and most species have wings with five veins (and other species outside of that group also have five veins) but one group has wings with three veins, the latter would be the shared derived feature uniting those species. Hennig also argued that shared ancestral character states

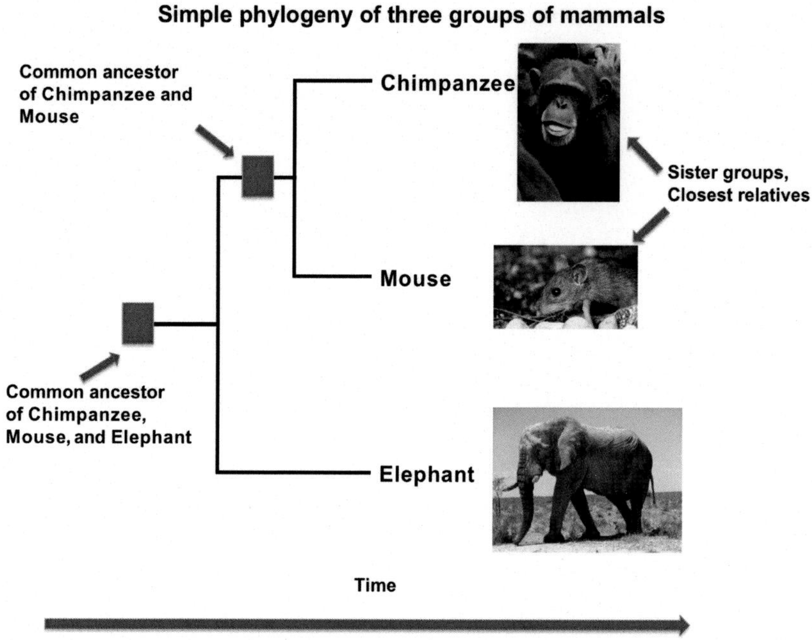

Figure 2.8 Simple tree of relationships, using mammals as an example.

should not be used to build phylogenetic trees because these did not provide the best insights into how organisms were actually related evolutionarily (e.g., the five-veined state above). Hennig proposed the use of a close relative of the group under investigation to determine the ancestral vs. derived state of a character. He also was aware that some character states could evolve in *parallel* in different groups or be subject to *reversal* and that these could complicate tree building. As an example, a character state present in an ancestor (say, red flowers) may experience an evolutionary change to white in its evolutionary descendants; however, one or more of these evolutionarily derived species could experience a reversal in that character state back to red.

TREE BUILDING 101

It is beyond the scope of this book to provide a detailed account of using characters to build trees (readers should consult Hall, 2011; Baum and Smith, 2012), but using a simple matrix of characters and states and the very basic approach that is summarized here, a simple phylogenetic tree is

Three imaginary species (A, B, and C) and an outgroup (O)

Outgroup	Ingroup

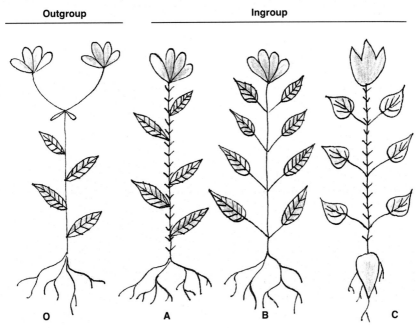

O   A   B   C

Figure 2.9 Tree-building tutorial. Three imaginary species and an outgroup. *Drawing courtesy of Elena Mavrodieva.*

not difficult to construct by hand for a small number of species and characters (see a brief tutorial in Fig. 2.9). First, one needs to assemble characters and states for those characters. As a simple example, consider the color of flower petals. Perhaps, there are several states in the group under study—red, white, or blue. Petal number is another character, with four and three petals as the observed states. Or, for a group of insects, wing venation might be a character with several states, three or five veins, as described in the example above. More and more characters and states could be added to a data set in this way. The example in Table 2.1 (modified from Judd et al., 2008) shows a simple scoring of characters for three species of imaginary plants. The characters used are simple—leaves are alternate or opposite; the stems are hairy or hairless; the petals are four or three.

How does an investigator determine what was the ancestral state and what is the derived state without a time machine where the ancestor could actually be observed? Scientists get around this problem by using

Table 2.1 States of morphological characters used in the analysis of the three imaginary species shown in Fig. 2.9

Morphological character	Character state[a]	
	Ancestral	Derived
1. Roots	Thin (0)	Tuber (thick) (1)
2. Stems	No hairs (0)	Hairs (1)
3. Leaves	Alternate (0)	Opposite (1)
4. Venation	Pinnate (0)	Palmate (1)
5. Petiole of leaf	Lacking (0)	Present (1)
6. Base of blade	Acute (0)	Cordate (1)
7. Flower parts	4 (0)	3 (1)
8. Flower parts	Separate (0)	Fused (1)
9. Flowers[b]	In a group of 2 (0)	Solitary (1)

[a]Character state codings are given in parentheses.
[b]Note that inflorescence condition (flowers solitary versus flowers in groups of 2) cannot be characterized further unless additional outgroups are used.

several close relatives of the group that is under investigation—these are called outgroups (Fig. 2.9). Those states shared by several outgroups are considered most likely to be ancestral. We can build a simple matrix with the state of the outgroups scored as 0 and the derived state as 1 (see the tutorial in Fig. 2.9; modified from Judd et al., 2008).

Of course, these characters will work only in very specific groups of organisms—flower color does not work with amphibians. Wing venation patterns do not work in fungi. What characters of morphology can be scored across all life? None! Put simply, one cannot build a tree of all life in this way. This limitation is just one of the reasons why the dual goal of elucidating evolutionary relationships among all species of living things and displaying those relationships in the form of an enormous phylogenetic tree emerged as one of the most daunting scientific challenges ever undertaken.

Therefore, with characters and the states of those characters assembled in a table, a tree can be built—but how to construct a tree from this information? Hennig's method involved building a separate tree for each character and then compiling these separate character trees into a summary. The problem is that as the numbers of characters and species increase, the trees become more complex and there may be multiple possible alternative summaries. How do we choose among them? The solution was to use the principle of *parsimony*. The term parsimony is from the Latin *parsimonia* (which means frugality or simplicity). Essentially, this principle means that the simplest solution is the best. Although the parsimony approach was used by the ancient Greeks for other purposes, the principle of parsimony in science was made famous by William of Occam (or Ockham; AD 1285—1349; Brown, 1990): "Plurality is not to be posited without necessity" (from Brown, 1990). This became known as Occam's razor because he used this approach to cut his opponents' arguments to shreds in scientific debates. As applied to the building of phylogenetic trees, the principle of parsimony means that the tree with the fewest number of character state changes (i.e., the simplest solution) is accepted as the best tree.

This approach of choosing among alternative trees is illustrated in Fig. 2.10. It is easy to examine the alternatives and determine the tree with the simplest interpretation of the data—that is, the fewest number of changes. The parsimony method is covered in more detail in Swofford et al. (1996) and in Felsenstein (1978). Although evolution does not always proceed in a parsimonious manner, this methodological approach uses the criterion of parsimony to choose the "best" interpretation of the data.

This has been just a simple overview of the history of methods for building phylogenetic trees. The entire book is devoted to this complex topic, reviewing many methods for both tree-building and assessing how much confidence one should have in a constructed tree. Considerable time and effort have been devoted by clever researchers to develop new and faster methods of building trees. Covering these in detail is beyond our scope (although some of the methodological challenges in building big trees are covered a bit more in Chapter 3: Building the Tree of Life: A Biodiversity Moonshot). Other common methods of building phylogenetic trees that we will not discuss here include Maximum Likelihood, Bayesian Inference, and Neighbor Joining (reviewed in Stuessy et al., 2014). By the 1980s and 1990s, tree-building had become a full-fledged science! Yet, assembling an actual tree of all named life remained unfathomable—a biodiversity moonshot.

READING PHYLOGENETIC TREES

Learning how to read and to interpret a phylogenetic tree requires some practice. To some people, understanding the meaning of a phylogenetic tree is intuitive because of family trees and the branching patterns they depict. Phylogenetic trees of species relationships can be viewed in the

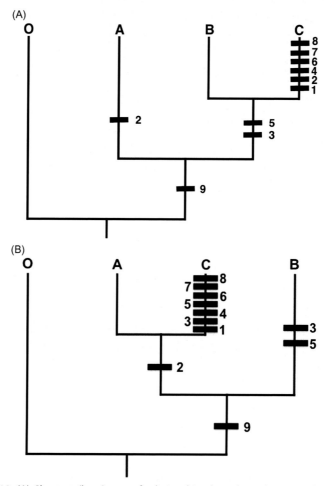

Figure 2.10 (A) Shortest (best) tree of relationships based on characters in Table 2.1 and Fig. 2.9. (B) Another tree (longer) of relationships based on the same characters in Table 2.1 and Fig. 2.9.

same way. Species occupy the end points or "tips" of the tree. Closely related species are connected by branches, and closely related groups of species (in some cases, combined into genera) are then connected deeper in the tree and so forth.

But these branching relationships can be shown in diverse formats, and this diversity may be initially confusing to the novice. Sometimes trees are drawn using a vertical format with the names at the top of the tree and the branches extending vertically from a common ancestor

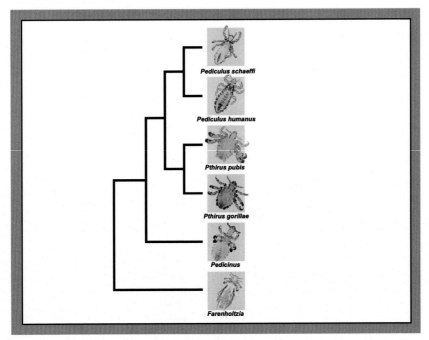

Figure 2.11 Different ways of depicting trees—horizontal. *Courtesy of David Reed, Florida Museum of Natural History, University of Florida.*

(Fig. 2.10). Darwin's simple tree from *On the Origin of Species* is drawn in this format (Fig. 2.1). Many readers like this tree format because it is similar to the human family tree diagrams with which we are so familiar, although in a typical human family tree the patriarch and matriarch are at the top and the descendants are below; this is the opposite orientation of Darwin's figure. Another common style is to use a horizontal tree format with the species names to the side (Fig. 2.11). Still another format is the circle tree with the ancestral branch in the very center and the species at the tips of the branches at the outer ring of the tree; a circle tree is a good way to show a tree of many species in a relatively small space, and these are now commonly used (Fig. 2.12). The formats for illustrating trees are all interchangeable. A horizontal tree format could be changed to a circle tree or vice versa; the relationships depicted would still be the same, only the format would change. Despite this diversity of approaches, visualization methods for capturing the beauty and complexity of large phylogenetic trees, in particular, require further development by

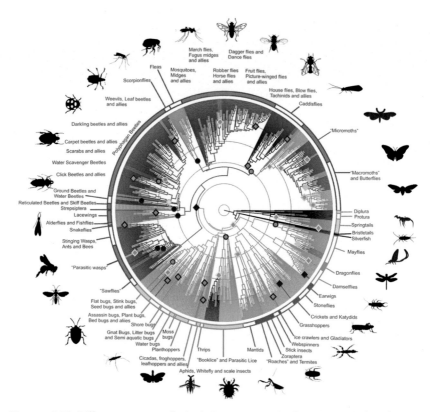

Figure 2.12 Different ways of depicting trees—circular. *From Wikipedia Free Commons.*

computer scientists and graphic designers, which is an active, challenging, and ongoing area of research (see Chapter 7).

REFERENCES

Baum, D.A., Smith, S.D., 2012. Tree Thinking: An Introduction to Phylogenetic Biology. Roberts and Co., Greenwood Village, CO.

Bessey, C.E., 1915. The phylogenetic taxonomy of flowering plants. Ann. Mo. Bot. Gard. 2, 109–164.

Brown, S.F. (Ed.), 1990. Philosophical Writings. A Selection. William of Ockham. Hackett, Indianapolis, IN.

Cronquist, A., 1968. The Evolution and Classification of Flowering Plants. Houghton Mifflin, Boston, MA.

Cronquist, A., 1981. An Integrated System of Classification of Flowering Plants. Columbia University Press, New York.

Darwin, C., 1859. The Origin of Species: By Means of Natural Selection, Or the Preservation of Favoured Races in the Struggle for Life. Cambridge University Press, Cambridge, UK.

Donoghue, M.J., Kaderiet, J.W., 1992. Walter Zimmerman and the growth of phylogenetic theory. Syst. Biol. 41, 74–85.

Felsenstein, J., 1978. The number of evolutionary trees. Syst. Zool. 27, 27–33.

Haeckel E., 1866. Generelle Morphologie der Organismen. Vols I and II. Georg Reimer, Berlin.

Hall, B.G., 2011. Phylogenetic Trees Made Easy: A How to Manual. Sinauer Associates, Sunderland, MA.

Hennig, W., 1950. Grundzüge einer Theorie der phylogenetischen Systematik. Deutscher Zentralverlag, Berlin, Germany.

Hennig, W., 1966. Phylogenetic Systematics. University of Illinois Press, Urbana, IL.

Judd, W.S., Campbell, C.S., Kellogg, E.A., Stevens, P.F., Donoghue, M.J., 2008. Plant Systematics: A Phylogenetic Approach, third ed. Sinauer Associates, Sunderland, MA.

Stevens, P.F., 1983. Augustin Augier's "Arbre Botanique" (1801), a remarkable early botanical representation of the natural system. Taxon 32, 203–211.

Stuessy, T., Crawford, D.J., Soltis, D.E., Soltis, P., 2014. Plant Systematics: The Origin, Interpretation, and Ordering of Plant Biodiversity. Regnum Vegetabile 156.

Swofford, D.L., Olsen, G.J., Waddell, P.J., Hillis, D.M., 1996. Phylogenetic inference. In: Hillis, D.M., Moritz, D., Mable, B.K. (Eds.), Molecular Systematics. Sinauer Associates, Sunderland, MA, pp. 407–514.

Wagner Jr., W.H., 1980. Origin and philosophy of the groundplan-divergence method of cladistics. Syst. Bot. 5, 173–193.

FURTHER READING

Farris, J.S., 1970. Methods for computing Wagner trees. Syst. Zool. 19, 83–92.

Hillis, D.M., 1996. Inferring complex phylogenies. Nature 383, 130–131.

Lamarck, J.B., 1809. Philosophie Zoologique. Museum d'Histoire Naturelle(Jardin des Plantes), Paris, France.

Swofford, D.L., 1998. PAUP* 4.0: Phylogenetic Analysis Using Parsimony (and Other methods), Beta Version 4.0. Sinauer Associates, Sunderland, MA.

Swofford, D.L., 2002. PAUP*. Phylogenetic Analysis Using Parsimony (*and Other Methods), Version 4. Sinauer Associates, Sunderland, MA.

Swofford, D.L., Begle, D.P., 1993. PAUP Version 3.1 User's Manual. Liiniois Natural History Survey, Champaign, IL.

Willson Stephen, J., 1999. Building phylogenetic trees from quartets by using local inconsistency measures. Mol. Biol. Evol. 16 (5), 685–693.

CHAPTER 3

Building the Tree of Life: A Biodiversity Moonshot

We choose to go to the Moon in this decade and do the other things, not because they are easy, but because they are hard; because that goal will serve to organize and measure the best of our energies and skills, because that challenge is one that we are willing to accept, one we are unwilling to postpone, and one we intend to win.

John F. Kennedy, speaking at Rice University, September 12, 1962

THE TREE OF LIFE: WHAT TOOK SO LONG?

Despite the progress in the approaches for building trees of relationships in the late 1900s, most trees that were constructed during that time were relatively small—fewer than 50 or 100 species. By 1993, a tree of 500 species was published, representing a landmark event given that only a few years earlier large trees were considered impossible to build (Chase et al., 1993); species with over 1000 species followed by the early 2000s. But it was not until 2015 that the first rough draft tree of all named life (2.3 million species) was finally published (Hinchliff et al., 2015). If Darwin, who began sketching trees of relationship long before he referred to the "Great Tree of Life" in *On the Origin of Species* were alive today, he would likely ask "what took so long?" when made aware of the recent publication of the first tree of all named life. Why was building the first rough draft Tree of Life so difficult?

While the answer has several components, the simple explanation is that tree building is hard, and building really big trees across all of life is extremely difficult. Building big trees of relationships required a data source other than appearance (morphology), and the use of DNA sequence data revolutionized tree building. Tree building is also extremely challenging computationally, and building enormous trees of thousands of species rivals even the most difficult analytical problems in physics, math, and astronomy. The invention of computational tools and computer

The Great Tree of Life
DOI: https://doi.org/10.1016/B978-0-12-812553-3.00003-5

power were therefore also essential. For several reasons, building the Tree of Life has long been considered a grand challenge in biology, a moon-shot for biodiversity. This is a challenge that could not be met until very recently, and it required the perfect storm.

THE PERFECT STORM

Building the Tree of Life was only made possible in the last decade via a perfect storm of algorithm development, DNA sequence data, and computer power. These breakthroughs made it possible for a team of computer scientists and organismal biologists to build the first rough draft Tree of Life for all of the 2.3 million named species on planet Earth! And the emphasis has to be on "team." In a sense, the story of how this cooperation came to pass represents much of what is best about modern science—teams of investigators with different areas of expertise work together to solve major problems that individuals representing separate disciplines could not conquer alone.

A key component of the perfect storm was DNA sequence information. DNA sequence data quickly became the fundamental tool of choice—the major line of evidence for building trees once it became possible to sequence genes easily and quickly. DNA sequence data are the perfect tree-building tool. With features of morphology, traits and states can be difficult to ascertain and score (see Fig. 2.9), but with DNA sequence data it is relatively easy—genes can be compared, and G, A, T, and C, as the building blocks of DNA, are the obvious states. And there are thousands of genes to sequence and compare across organisms, each with hundreds of characters (each consisting of either G, A, T, C) to compare—a gold mine of useful characters for building trees (Fig. 3.1). Early breakthrough studies in the 1990s on plants revealed that it was possible, albeit difficult at that time with the computer power and the gene sequences then available, to build trees of relationships for 500 species (Chase et al., 1993). The floodgates were open.

But the more and more DNA data there are to analyze, the more important it is to have the computer power to analyze massive amounts of data for numerous species. Also needed were improved algorithms and methods that make it possible to streamline the tree-building analytical process. Thus, the perfect storm involved improvements not just in sequencing technology but also in greater computer power and computational methods. And, of course, the organismal biologists to collect and

Figure 3.1 DNA sequence matrix showing the alignment of nucleotides. It shows DNA sequences of the same gene from different species aligned and color coded by nucleotide. The bacterium (second entry on left) has a unique stretch of nucleotides near the left side—to align the region with the remaining sequences, spaces (−) were introduced manually (or by computer) in all remaining species. *From Wikipedia Free Commons.*

identify samples are critical—without field biologists, biodiversity would not be correctly identified, collected, and catalogued. Actually, the modern biodiversity scientist has to do a bit of all of these things—in layman's terms, the modern biodiversity scientist is sort of the decathlete of science.

THE TREE OF LIFE: TOO BIG TO NAIL

Building big trees of relationships is far more complex than is often realized in that the number of possible trees that can be produced for even a small number of species is staggering. Thirty years ago, when our lab and others started building "big trees" of just a few hundred species, we were told this was an impossible task—some simple math reveals why. Felsenstein (1978) showed that the number of possible trees actually increases exponentially as you add species. For 4 species, there are 15 possible rooted trees, but for just 10 species, there are 282 million rooted trees; by the time you reach 22 species, the number of possible trees is 3 $\times$ 10^{23}, which approximates Avogadro's number (6.02 $\times$ 10^{23}), which you may remember from chemistry is the number of particles in a "mole." Hence, with a mere 22 species, there are roughly a mole of possible trees. In the vicinity of 228 species, Hillis (1996) estimates that the number of possible trees that can be assembled exceeds the number of atoms in the universe (Fig. 3.2).

It should be clear based just on the number of possible solutions (trees) reviewed above that building and evaluating large trees that include thousands to millions of species is really challenging. As a result, the dual goals of elucidating evolutionary relationships among all species of living things

No. Species	Number of possible rooted trees
1	1
2	1
3	3
4	15
5	105
6	945
7	10,395
8	135,135
9	2,027,025
10	34,459,425
11	654,729,075
12	13,749,310,575
13	316,234,143,225
14	7,905,853,580,625
15	213,458,046,676,875
16	6,190,283,353,629,375
17	191,898,783,962,510,625
18	6,332,659,870,762,850,625
19	221,643,095,476,699,771,875
20	8,200,794,532,637,891,559,375

Figure 3.2 The number of possible trees increases exponentially as you add species. *Redrawn from Felsenstein (1978).*

and displaying those relationships in the form of an enormous phylogenetic Tree of Life therefore emerged as one of the most profound and daunting scientific challenges ever undertaken. With more than 2.3 million species described, and many more millions undiscovered or extinct, the size of the Tree of Life is immense. No wonder, then, that assembling the Tree of Life has long been considered a biodiversity moonshot (reviewed in Soltis et al., 2010), a task that many considered impossible (Graur et al., 1996; Willson, 1999). Many researchers advocated that large trees of relationship could not really be constructed—tree building had to focus on small numbers of species so that the tree-building analyses could be thorough and completed in a reasonable time frame. Some researchers thought it was imperative to break up the problem of building phylogenies into a number of smaller problems of four species at a time (Willson, 1999). But it was unclear how this approach of "going small" could possibly work with any large group of organisms—how to build a tree for 10,000 birds or 350,000 flowering plants or the millions of species in the Tree of Life?

For young people growing up in the 1960s and 1970s, the words of a youthful president who said we could do the impossible were deeply influential. Kennedy's moonshot speech has long been inspirational to biologists—why is something considered impossible? Maybe there is a solution. Those words instilled at an early age that we as individuals should strive to tackle the most difficult and daunting problems, and that with determination and cooperation (as well as good timing and luck!) many difficult scientific problems can be surmounted.

In addition to the challenges posed by the enormous scale of tree building, obtaining data across the Tree of Life is also a challenge. As reviewed in Chapter 2, in building trees, the investigator assembles a matrix of characters and character states—the modern approach is to rely on DNA sequence data. But despite the power of DNA data, a major problem is that there are very few genes shared across all of the enormous diversity of life. Estimates are that the number of such genes may be approximately 87 genes (https://www.ncbi.nlm.nih.gov/books/NBK26866/#_A61_). Keep in mind that bacteria and archaea typically have 1000−3000 genes; eukaryotes typically have many more, in the range of 15,000−25,000 (there are 21,000 genes in *Homo sapiens*). Only 87 shared genes is a fairly small number, making it problematic to construct a comprehensive Tree of Life.

Some of the profound early insights into the major groups of life relied on just a few genes that were shared among all species of life. Based on two ribosomal RNA genes, genes that code for ribosomes (location of protein synthesis in the cell) and are fundamental for life, it was found that life could be placed in three distinct groups, two of which are microbial (archaea and bacteria) and the eukaryotes (Woese and Fox, 1977). This view of the major groups of life has since been revised by some to perhaps just two major groups, as will be reviewed in Chapter 4, How Many Living Species? However, it is also important to realize that these two genes, as well as the other genes shared across life, evolve too slowly to resolve relationships to the species level across all of life. Thus, while getting a well-resolved tree to the level of species with these genes is not feasible, these genes can provide insights into the major groups and clarify the backbone of the Tree of Life.

The past two decades have seen extensive tree building on specific groups of organisms, thanks in large part to DNA sequencing. For example, early studies of green plants paved the way by sequencing a gene fundamental to photosynthesis (see Ritland and Clegg, 1987). Analyses of mammals, birds, and other vertebrates relied on genes found to be useful in resolving relationships in those groups of organisms. The fungal biologists used other genes in their tree building. The DNA sequence revolution in the study of biodiversity was in full swing! The result has been the most dramatic change in our understanding of relationships and evolution in many, perhaps most, groups of organisms in the past several hundred years. The DNA sequencing and tree-building revolutions had ushered in what will be remembered as a Golden Era of our understanding of the relationships of biodiversity.

But despite years of building trees for subgroups of life and numerous new insights for specific groups, a comprehensive tree of all life remained a daunting task that seemed out of reach until just a few years ago. One challenge is that most researchers focused on their own groups of interest—say, birds, snakes, flowering plants, fungi—and there was little emphasis on synthesis. How to combine sequence data across all of these groups of life when different genes were used across different parts of the Tree of Life? A matrix of the many genes and species studied could be put together—this is termed a "supermatrix" (de Queiroz and Gatesy, 2007). Supermatrices have grown larger and larger with the ease of DNA sequencing. Pipelines are now used to construct data sets for thousands of species for several genes and for numerous genes and hundreds of species. Data sets will only get larger, encompassing more and more genes as sequencing technology advances. However, the lack of significant overlap in data across all of life made this approach problematic in building the Tree of Life.

Another approach used to build more and more comprehensive phylogenetic trees is to combine smaller phylogenetic trees into a single, larger comprehensive tree (e.g. Bininda-Emonds 2004). This is termed a supertree approach. In contrast to supermatrices, which combine underlying data, supertree methods take groups of trees with at least some overlapping species and construct a comprehensive tree, called a supertree, from these individual input trees. The supertree contains all of the species present in the input trees, and because it combines trees that overlap in content, it can contain more species than any of the input trees (Fig. 3.3). Supertrees and approaches used to build supertrees are well-reviewed (e.g., Sanderson et al., 1998; Bininda-Emonds and Sanderson, 2001; Gordon, 1986). They have been used to provide impressive trees for some groups of life, including birds (Sibly et al., 2012) and mammals (Bininda-Emonds et al., 2007).

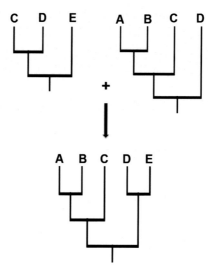

Figure 3.3 Building what is termed a "supertree" by melding or gluing trees together that have some overlap in the species (labeled A—E) they contain.

But there are also problems with supertees. Supertree methods are not explicitly targeted toward the exploration of variability among the input trees—that is, if two trees published by different teams of researchers for the same set of organisms differ in branching pattern, or different gene sequences tell slightly different stories of relationships for that group, supertree methods have problems. Supertrees also have difficulty when the number of shared species between two sets of trees is low. Also, supertree methods can sometimes generate relationships not found in any of the original smaller trees (the group D + E in Fig. 3.3). As a final problem, there is currently no supertree method that scales to building the entire Tree of Life. "Houston, we have a problem." A second revolution in tree-building methods was needed to achieve a biodiversity moonshot—one that would enable a giant leap in producing trees of relationships.

GRAPH THEORY, GOOGLE, AND BUILDING THE FIRST TREE OF LIFE

Phylogenetic trees are the most commonly used way to show patterns of relationships. But, as reviewed above, there are problems when attempting to build massive trees on the scale of all, or even large swaths, of life. However, trees of relationship as so far discussed can be problematic when attempting to summarize multiple trees. Working with trees of relationship can be especially difficult when trying to accommodate conflict between multiple data sets and biological processes such as horizontal gene transfer (HGT) and hybridization (see below). Also problematic is combining partially overlapping sets of trees—trees that share just some species. How to overcome these problems to build a tree of all named life?

One way to solve the problem of building really big trees of thousands or millions of species is to use graph theory (Smith et al., 2013). To review a little math (don't panic!), graph theory is defined as "the study of graphs, which are mathematical structures used to model pairwise relations between objects." In this sense, a graph consists of points connected by lines (a simple graph is shown in Fig. 3.4). There is a lot more to graph theory than this (see Trudeau, 1993, for review). But you know much more about graph theory than you realize—your life is immersed in graph theory. Graph theory is what Google and others use to connect the dots of your life: for example, the purchasing habits of an individual. What are the places you visit when you shop online? Who is most likely to visit Target or Pier One or Bath and Body Works? Connect the dots. Now connect the dots over millions of consumers and search for repeating patterns or connections. These are big complex problems, and solving them employs graph theory. Consider the examples below.

If you use social media, consider the numerous connections you have and how those are represented (Fig. 3.5). The field of study called social network analysis involves the "mapping and measuring of relationships and flows between people, groups, organizations, computers, URLs, and other connected information/knowledge entities." The nodes in the network are the people, and the links reveal relationships between the nodes. Fig. 3.5 is a graph that illustrates those connections and effectively depicts the personal relations of Internet users. The social networks of those on social media are graphs that are based on graph theory. Graph theory is responsible for suggesting whom to follow on Instagram and whom to friend on Facebook.

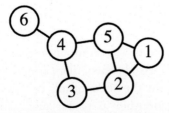

Figure 3.4 A simple graph. *From Wikipedia Free Commons.*

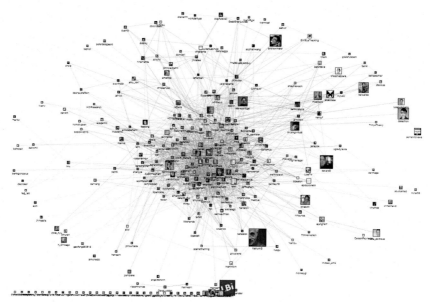

Figure 3.5 Social graph of connections among people. *From Wikipedia Free Commons. Source: https://www.flickr.com/photos/marc_smith/4511844243.*

A major use of graph theory is for companies to investigate the connections among customers and by doing so more effectively market a product. These graphs of connectivity are massive and complex, but through graph theory they can be analyzed rapidly to ascertain possible new customers. For example, the spread of information (a sale on a product) via social media could reach new customers, and graphs can help ascertain the best approaches and most likely new customers.

Google Maps is another great illustration of the power of graph theory. You have likely used this technology on your phone to determine the best route from one location to the next. Google Maps can show you several alternative roots and indicate the fastest. And these calculations can be made rapidly, what appears to be almost instantly on your phone (Fig. 3.6).

So why not apply graph theory to building the Tree of Life? The speed and ability to deal with millions of data points are crucial central features of graph theory that could make it possible to compute the Tree of Life! This is, in fact, what Smith et al. (2013) proposed. Researchers had recognized that trees were already graphs (directed acyclic graphs to be precise) (Fig. 3.7). Smith et al. (2013) suggested that, perhaps,

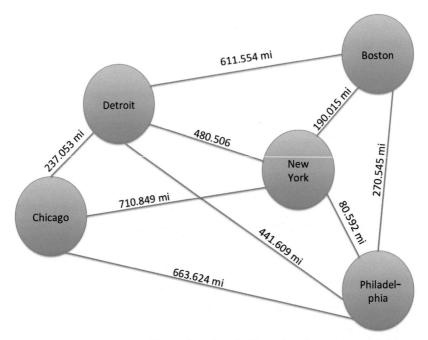

Figure 3.6 Graphs, as might be used in Google Maps showing routes and distance between points.

Moving from trees to graphs

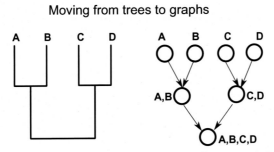

Figure 3.7 A tree of relationships (left) can be expressed as a graph (right).

researchers might find solutions in graph theory. Connecting millions of points (in this case species) can be accomplished by using graph theory and looking for consistent underlying patterns in connecting the points. Graphs also permit representation of conflict and uncertainty. We can also map new trees into the graph so that the overall Tree of Life can be updated. Graphs also allow the easy serving of data over a network to the community—much more easily than do trees.

To address the various problems noted above with trees, Smith et al. (2013) used graph theory and introduced new methods for aligning, synthesizing, and analyzing trees of relationship within a graph, which they called a tree alignment graph (TAG). The TAG can be queried and analyzed to accomplish the goals above: explore conflict and uncertainty among trees and map new trees into the graph to build and update the Tree of Life. A small portion of the flowering plant tree is shown as a graph in Fig. 3.8. So, basically, one dilemma of how to build the Tree of Life was solved by Smith et al. (2013) who established this graph-based method. Hinchliff et al. (2015) used those graph methods to build the first tree of all named life, represented here in the more familiar form of a tree (Fig. 3.9). However, even this solution presented new challenges due to complexities in missing data and other problems. These new challenges spawned innovative research into solving these complex phylogenetic problems that continue to be applied to building the Tree of Life (Redelings and Holder, 2017).

MORE CHALLENGES: WHAT'S IN A NAME?

Stitching trees together is also difficult, even without considering the complexities of graph theory. The trees (converted to graphs) have to have some overlap in species to have a match for connectivity. But the names of the species in the different trees have to match as well. The famous line of Shakespeare from Romeo and Juliet is applicable here: When Juliet says to Romeo, "What's in a name? That which we call a rose by any other name would smell as sweet." Bottom line is that names matter. What if the overlap between two trees (now graphs) involved different spellings of the same species … one of which was incorrectly spelled? Using the rose metaphor, let's say that one study used the correct name, *Rosa palustris*, but in another study, it was misspelled as *Rosa palustrus*. These names would not be considered the same any more than the names Katie and Katy or Sarah and Sara. Or perhaps one author only used the genus (*Quercus*) name, and the author of another paper used the entire species name (*Quercus alba*). Or one paper used common names, like "human," while another published paper used the scientific name *Homo sapiens* (Fig. 3.10). Computer algorithms will not recognize any of these as matches. So, the trees will not be connected in the graph database. Now imagine this problem on a massive scale of millions of names and species. Another problem encountered is that some plants and

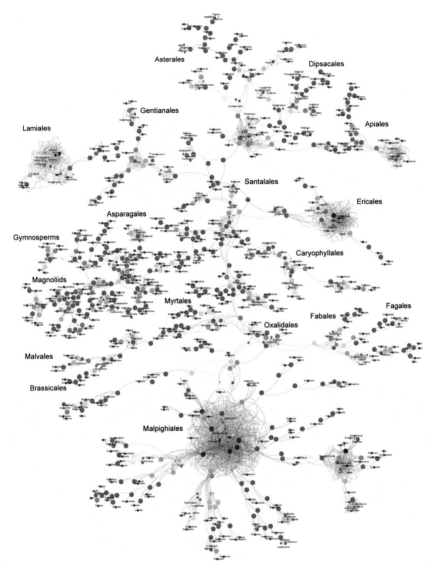

Figure 3.8 Flowering plant relationships expressed as a graph. *From fig. 2 of Smith S.A., Brown J.W., Hinchliff C.E., 2013. Analyzing and synthesizing phylogenies using tree alignment graphs. PLoS Comput. Biol. 9 (9), e1003223.*

animals were actually given the same scientific name. In other cases, some species have more than one given scientific name although only one name is correct. Determining the correct name can be a detective case rivaling your favorite murder mystery. Computer databases can help with

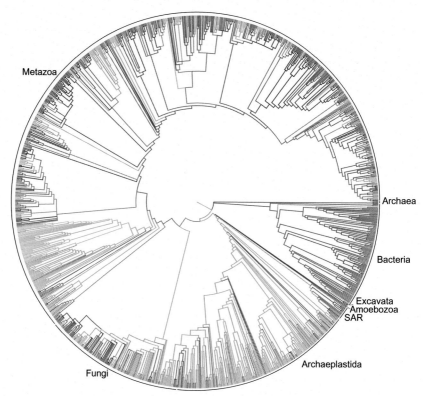

Figure 3.9 The first comprehensive tree of all named life. Colors represent the amount of DNA data. Red represents high amounts, blue represents no DNA data, and shades in between represent various amounts of DNA. Only 17% of named species have any DNA sequence data available for use in the first Tree of Life. *From Hinchliff et al., 2015. Synthesis of phylogeny and taxonomy into a Comprehensive Tree of Life. Proc. Natl. Acad. Sci. 112, 12764–12769.*

name recognition (e.g., *Rosa alba* variety *palustris* = *Rosa palustris*), but these have to be constructed across all groups of organisms. In many cases the name issues still have to be resolved manually.

TREE OF LIFE VERSUS NETWORK OF LIFE

One of the more intriguing discoveries of the genomic era of research in which we now live is the amazing exchanges of genes between organisms. We have known for a long time that closely related species can exchange genetic material via hybridization, but DNA data revealed that it was even more common than previously thought (Soltis and Soltis, 2009).

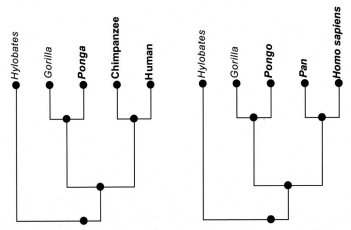

Figure 3.10 Names matter in building trees. The two trees shown of primates are identical, but different names are given for the same species. For example, human (in the tree on the left) and *Homo sapiens* (in the tree on the right) represent the same species (ours). Similarly, *Pan* and chimpanzee are names (scientific and common) for the same genus. However, the use of these different names results in problems in combining trees (building supertrees or networks of trees), because a computer algorithm will only consider identical names as matches. Even typographical errors (*Ponga* vs. *Pongo*; the latter is correct) will result in mismatches and errors in combining data and trees. Hence, tree building is challenging!.

If hybrids cross back to one of the parents, it can lead to a trickle of genetic material from one species to another. Ancient hybridization is suspected between our own species and Neanderthal.

However, there are additional levels of complexity to evolution. Related plants (and some animals) may experience hybridization followed by genome doubling. This results instantly in a new species that has two genomes—branches of life have come back together (Fig. 3.11). Life is full of these reticulation events, making it hard to draw the Tree of Life as a simple branching pattern of diverging branches—branches can come back together.

Perhaps even more intriguing is the exchange of genes between distantly related organisms that do not hybridize. For example, a number of parasitic plants have genes from their host plants (the plant species on which they live and from which they obtain nutrients). The parasite *Rafflesia* (which produces the world's largest flower) has genes in its genome from its host, which is a vine in the grape family. These are not closely related plants. This process is called horizontal gene transfer (HGT) or lateral gene transfer (LGT). This is the movement of genetic

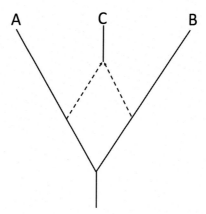

Figure 3.11 Hybridization between species is a widespread biological process—it is very common in many parts of the Tree of Life. Hybridization can be challenging to depict in trees, however, because it results in the merger of lineages (as shown here), rather than the conventional splitting of lineages typically depicted in trees.

material to an organism other than by the ("vertical") transmission of DNA from parent to offspring (Keeling and Palmer, 2008). LGT is an important factor in the evolution of many organisms. In this instance of *Rafflesia* and host, the close intimate contact of tissues from the host and parasite must have permitted genes to move from one species to another.

However, other examples of LGT are similarly remarkable, but harder to explain. A light receptor gene has moved from a moss relative called a hornwort to a fern (Li et al., 2014). The mitochondrial genome of the flowering plant *Amborella* has genetic material of mosses and other flowering plants (Rice et al., 2013). How did it get there?

LGT is most prevalent in the bacterial world where a high percentage of genes from one bacterium may have their source in other distantly related bacteria. The process of LGT may be so extensive in the microbial (bacteria and archaea) world that it becomes hard to represent relationships as a simple branching pattern. In reality, the bacterial Tree of Life may be more net-like (Dagan and Martin, 2009) (Fig. 3.12).

MILES TO GO BEFORE WE SLEEP

Although we now have a first draft Tree of Life (Hinchliff et al., 2015), it is highly preliminary. Many relationships among species in the tree remain poorly understood. This tree is simply a starting point. The goal now is to improve the tree—for scientists (with the help of the public) to work

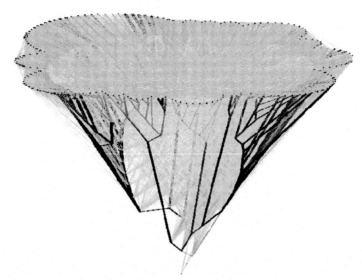

Figure 3.12 Another natural biological process that is important to consider is HGT between different lineages. Like hybridization, HGT will move genes from one species to the next. However, whereas hybridization occurs between closely related species, HGT can occur between distant species. HGT is known to be very common in the microbial world. As a result, the bacterial Tree of Life may be better shown as a net or web rather than a simple branching tree. *From fig 2 from Dagan, T., Martin, W., 2009. Getting a better picture of microbial evolution en route to a network of genomes. Phil. Trans. Royal Soc. B. 364, 2187–2196.*

on segments for which they have expertise and update and improve segments of the tree. It takes a village to build and improve the Tree of Life.

One major problem with the Tree of Life that we now have in hand is that it is based on very little data. DNA sequence data represent the backbone or the basic building material for constructing trees of relationships. Yet, we only have DNA data for perhaps 17% of the species on this tree. This can be seen clearly by looking at the colors in the simplified Tree of Life in Fig. 3.9. Note the colors ranging from red to blue. Red indicates where we have DNA data to support the tree (the archaea and bacteria are only placed using DNA data, but these represent only a small fraction of archaea and bacteria diversity on Earth), blue is where we do not have DNA data, and shades in between indicate varying amounts of DNA data. You can see the tree is mostly blue. Another problem discussed in Chapter 4 is that we have only described a fraction of life on our planet. Although specialists have named 2.3 million species, there are many millions still unnamed. We have a long way to go to understand the species of life on Earth.

REFERENCES

Bininda-Emonds, O.R.P., 2004. The evolution of supertrees. Trends Ecol. Evol. 19 (6), 315–322.

Bininda-Emonds, O.R.P., Sanderson, M.J., 2001. Assessment of the accuracy of matrix representation with parsimony supertree reconstruction. Syst. Biol. 50, 565–579.

Bininda-Emonds, O.R.P., Cardillo, M., Jones, K.E., MacPhee, R.D.E., Beck, R.M.D., Grenyer, R., et al., 2007. The delayed rise of present-day mammals. Nature 446, 507–512.

Chase, M.W., Soltis, D.E., Olmstead, R.G., Morgan, D., Les, D.H., Mishler, B.D., et al., 1993. Phylogenetic relationships among seed plants based on *rbc*L sequence data. Ann. Mo. Bot. Gard. 80, 528–580.

de Queiroz, A., Gatesy, J., 2007. The supermatrix approach to systematics. Trends Ecol Evol 22, 34–41.

Felsenstein, J., 1978. The number of evolutionary trees. Syst. Zoo 27, 27–33.

Li, F.-W., Villarreal, J.C., Kelly, S., Rothfels, C.J., Melkonian, M., Frangedakis, E., et al., 2014. Horizontal gene transfer of a chimeric photoreceptor, neochrome, from hornworts to ferns. Proc. Natl. Acad. Sci. U.S.A. 111 (18), 6672–6677. Available from: https://doi.org/10.1073/pnas.1319929111.

Gordon, A.D., 1986. Consensus supertrees: the synthesis of rooted trees containing overlapping sets of labeled leaves. J. Classification 3, 31–39.

Graur, D., Duret, L., Gouy, M., 1996. Phylogenetic position of the order Lagomorpha (rabbits, hares and allies). Nature 379, 656–658.

Hinchliff, C.E., Smith, S.A., Allman, J.F., Burleigh, J.G., Chaudhary, R., Coghill, L.M., et al., 2015. Synthesis of phylogeny and taxonomy into a Comprehensive tree of life. Proc. National Acad. Sci. 112, 12764–12769.

Hillis, D.M., 1996. Inferring complex phylogenies. Nature 383, 130–131.

Keeling, P.J., Palmer, J.D., August 2008. Horizontal gene transfer in eukaryotic evolution. Nat. Rev. Genet. 9 (8), 605–618.

Redelings, B.D., Holder, M.H., 2017. A supertree pipeline for summarizing phylogenetic and taxonomic information for millions of species. PeerJ. Available from: https://doi.org/10.7717/peerj.3058.

Rice, D.W., Alverson, A.J., Richardson, A.O., Young, G.J., Sanchez-Puerta, M.V., Munzinger, J., et al., 2013. Horizontal transfer of entire genomes via mitochondrial fusion in the angiosperm *Amborella*. Science 342, 1468–1473.

Ritland, K., Clegg, M.T., 1987. Evolutionary analyses of plant DNA sequences. Amer. Naturalist 130, S74–S100.

Soltis, D.E., Moore, M.J., Burleigh, G., Bell, C.D., Soltis, P.S., 2010. Assembling the angiosperm tree of life: progress and future prospects. Ann. Missouri Bot. Garden 97, 514–526.

Soltis, P.S., Soltis, D.E., 2009. The role of hybridization in plant speciation. Annu. Rev. Plant. Biol. 60, 561–588.

Sanderson, M.J., Purvis, A., Henze, C., 1998. Phylogenetic supertrees: Assembling the trees of life. Trends. Ecol. Evol. 13, 105–109.

Trudeau, R.J., 1993. Introduction to Graph Theory. Dover Publications, Inc, New York.

Sibly, R.M., Witt, C.C., Wright, N.A., Venditti, C., Jetz, W., Brown, J.H., 2012. Energetics, lifestyle, and reproduction in birds. Proc. Natl. Acad. Sci. U.S.A. 109, 10937–10941.

Smith, S.A., Brown, J.W., Hinchliff, C.E., 2013. Analyzing and synthesizing phylogenies using tree alignment graphs. PLoS Comput. Biol. 9 (9), e1003223.

Willson, S.J., 1999. Building phylogenetic trees from quartets by using local inconsistency measures. Mol. Biol. Evol. 16 (5), 685–693.

Woese, C.R., Fox, G.E., 1977. Phylogenetic structure of the prokaryotic domain: the primary kingdoms. Proc Natl Acad Sci U S A 74, 5088–5090.

CHAPTER 4

How Many Living Species?

It is entirely possible that specialists have discovered only 20 percent, or fewer, of Earth's biodiversity at the species level. Scientists working on biodiversity are in a race to find as many of the surviving species as possible ... before they vanish and thus are not only overlooked but never to be known.

E.O. Wilson 2016

HOW MANY SPECIES HAVE BEEN NAMED?

You can pull your cell phone out of your pocket or bag and determine exactly where you are on the surface of the Earth, precisely how far it is to the nearest coffee shop, and how many steps you have taken today. But, you will not be able to use your phone or any other device to tell you how many species there are on our planet because we simply do not know. You will only be able to obtain various approximations—and these estimates of the number of species of life on Earth are uniformly large.

The best estimate of the number of named species is 2.3 million. We stress "estimate" again because we are not even sure exactly how many species scientists have already named. That may seem odd to not only the nonspecialist, but also scientists in other areas of research. How could we not know what species scientists have named so far? There are several reasons for this lack of clarity. Sometimes the same species was named independently by different scientists who gave the species different names, but did not know of the other's work. Imagine two scientists in the 1800s, one from England and the other from the Netherlands, both independently traveling to collect in remote areas of China, and separately collecting and naming the same species—but using and publishing different names in different journals and never aware of the other's work. This was particularly problematic in the late 17th century to the early 19th century when communication of results was not as rapid and uniformly available as today.

The Great Tree of Life
DOI: https://doi.org/10.1016/B978-0-12-812553-3.00004-7

There are other reasons why we do not know precisely how many species have been named. Perhaps one scientist decided to include a previously named species within another species—essentially that species then vanished. But perhaps it really is a very distinct entity according to other biologists. Perhaps one scientist recognized two entities as varieties—not different enough to be considered species—but another specialist did consider them distinct enough to be species. These issues of naming species (the application of nomenclature) are often very difficult to resolve. This is not glamorous research—imagine going through a phone book (these are used less and less these days) of a large city looking for errors—tedious, underappreciated work, yet critical to the cataloging of life.

Parts Unknown

Although scientists have named roughly 2.3 million species, much of the living diversity on our planet remains undiscovered, undescribed, and hence unnamed and unknown. The estimates of how much life remains unknown to humans in the scientific sense (unnamed) vary, but are consistent in one sense—the estimates are very large, typically in the range of 5−20 million, with some as high as 50 million or more species (Costello et al., 2013; Scheffers et al., 2012). As scientists, including some of the most prominent naturalists of our time (E.O. Wilson, Peter Raven), have long stressed, there is actually much more that we do not know than what we do know about the diversity of life inhabiting our planet (Stork, 1993; Sangster and Luksenburg, 2015). As an example, consider those organisms we call eukaryotes, organisms in which the DNA is organized in structures called chromosomes and present in a nucleus. Eukaryotes include our own species and represent all life except the bacteria and archaea (sometimes also called archaebacteria; these are bacteria-like organisms not discovered until the 1970s by Carl Woese and collaborators). We may have discovered only 20% or less of all species of eukaryotes. Although perhaps 2.3 million eukaryotes have been named, there may be 10 million or more total species—others suggest that there may be 50 million eukaryotes. Furthermore, although there is more diversity on land than in the oceans (Dawson, 2012; Grosberg et al., 2012), what we do know about the eukaryote world is highly biased toward terrestrial environments—86% of living species on the surface of the Earth and 91% of species in the ocean still await description (Mora et al., 2011). For the microbial world (bacteria and archaea), our knowledge gap between the number of named species and the actual number of living species is much worse (see the "We Are Living in a Bacterial World" section).

Even for the parts of the Tree of Life where many species have been named, our knowledge of those species is typically limited. For example, for most species that have already been named, we have no DNA sequence data. Scientists have been collecting and naming species for hundreds of years, but it is really only in the past several decades that we have been able to start using DNA sequence data to determine how species are related to each other. DNA sequence information is the critical underlying data that provide the best means to place species with accuracy in the Tree of Life. However, with so many species on our planet and low funding for basic research, as well as limited scientific person-power to do the work, we still have only obtained DNA sequence data for a fraction of life on Earth. That said, with a relatively modest monetary investment—a major biodiversity initiative—significant progress could be made rapidly. Without DNA data, species are placed by name only and placed in an unresolved position in the Tree of Life. If an oak species (genus *Quercus*) has no sequence data, it is placed in the genus *Quercus*, but as a branch with no close oak relatives—because its closest relatives are not known.

The many species that are named, but have no DNA data, are sometimes referred to as dark parts of the Tree of Life. Amazingly, only 17% of named life has any snippets of DNA sequence data available—hence, much of the Tree of Life remains very dark. But this percentage is misleading in that some groups are well understood and have a lot of DNA data available (e.g., vertebrates), whereas most lineages of life are far more poorly understood and have much less coverage than 17% of the known species. The estimate of DNA data availability can also be misleading. Sometimes the DNA data deposited for a species are so limited or poor in quality that they are useless in building the Tree of Life.

Cryptic Species

A greatly underappreciated problem in discovering and naming the biodiversity on Earth is that there are large numbers of what are termed cryptic species (Scheffers et al., 2012). By cryptic we mean species that are very similar to another species in appearance and as a result were not initially discovered, that is, entities that were ultimately determined to be genetically distinct, but are morphologically very similar (Pérez-Ponce de León and Poulin, 2016). We tend to recognize and describe entities that obviously differ to the eye. If two entities differ only slightly in appearance, it was traditionally not considered important to recognize this difference by naming another species. Disentangling and recognizing that cryptic species exist takes careful research. Only detailed study by biologists has revealed the distinctive nature of entities that, because of their superficial similarity in appearance, had simply been lumped together as a single species in the past. For example, there are frogs that are very similar in appearance, but have distinctive vocalizations or calls and hence do not interbreed. This difference keeps them genetically distinct—they are therefore separate species—however, that was only discovered via careful research (see below).

Cryptic species are not a rarity on Earth. We likely have noticed only a small fraction of these. However, cryptic species may be more likely to be found in some groups of organisms than others (Pérez-Ponce de León and Poulin, 2016). Those who study Cnidarians (sea anemones, corals, and jellyfish) estimate that cryptic species may represent a huge, unrecognized aspect of diversity in the marine world. The data available suggest that, on average, each named species of Cnidarian recognized based on appearance (morphology) may actually represent two to five genetically distinct species. Similarly, cryptic species may be prevalent in the fungal world. We do not see obvious morphological differences between entities such as mushrooms and molds that nonetheless are genetically very different.

Cryptic species abound in the plant world, as well. Plants often produce novel capabilities and new species via a process called polyploidy, or genome doubling. Imagine a species with 14 chromosomes. A mutation yields offspring with 28 chromosomes—these plants would be immediately isolated reproductively from the parent. Plants seem to tolerate this process and even have thrived in terms of diversity because of the prevalence of genome doubling. But, these entities often go unrecognized because the plants with different chromosome numbers are so similar in appearance. As an example, we studied the flowering plant *Tolmiea*, the common piggy back plant in the western part of North America (Soltis, 1984; Soltis and Soltis, 1989); it is also grown as an ornamental with plastic replicas commonly found hanging in restaurants.

Since its discovery (over 170 years ago), *Tolmiea* had been considered to consist of a single species, *Tolmiea menziesii* (Fig. 4.1). Eventually, when

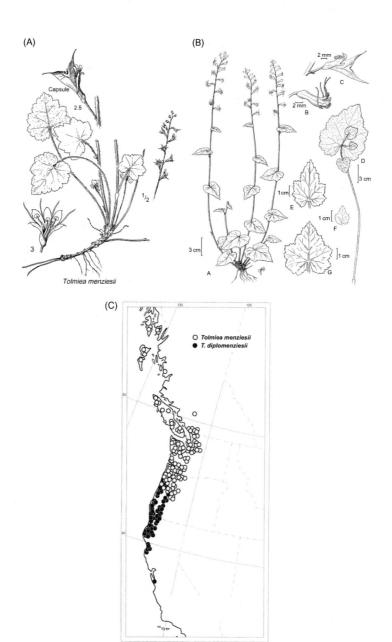

Figure 4.1 Cryptic species (species that look very similar, but are distinct genetically and often differ in habitat and distribution) are now known to be prevalent across the Tree of Life; here is an example in plants. (A) *Tolmiea menziesii*, line drawing. (B) *Tolmiea diplomenziesii*, line drawing. Note the similarity to *Tolmiea menziesii*. (C) Geographic distribution of the two species of *Tolmiea*; each species has a distinct range; they also differ in chromosome number (see text). *From (A) Hitchcock, C. L., Cronquist, A., 1961. Flora of the Pacific Northwest. In: Vascular Plants of the Pacific Northwest. Part 3. Saxifragaceae to Ericaceae. University of Washington Press, Seattle, pp. 68. (B and C) Judd, W.S., Soltis, D.E., Soltis, P.S., 2007.* Tolmiea diplomenziesii: *a new species from the Pacific Northwest and the diploid sister taxon of the autotetraploid* T. menziesii *(Saxifragaceae). Brittonia 59, 218−225.*

it became possible to count chromosomes, it was shown that this species had 28 chromosomes based on the collection of a plant from the state of Washington, U.S.A. However, a plant from California was later investigated, and it had only 14 chromosomes (Soltis, 1984). After more research, it was shown that populations in the south of the range have 14 chromosomes and those in the northern part have 28 chromosomes. Furthermore, there is no major geographic overlap of these two "races" or "cytotypes" and no zone of hybridization where they come into contact (Fig. 4.1). In addition, attempts to cross the two types failed—they are reproductively isolated. The two races differ genetically and have different ecological niches and also differ in terms of water use efficiency (Judd et al., 2007; Visger et al., 2017). These two races are therefore actually distinct, cryptic species that were not recognized until after careful study. Populations with 14 chromosomes were ultimately named a new species, *Tolmiea diplomenziesii* (Fig. 4.1; Judd et al., 2007); this is a great example of a cryptic species.

Chromosome races in plant species abound, and there are many more such cryptic species similar to the one in *Tolmiea* described here, awaiting discovery (Soltis, 1984; Soltis et al., 2007). This example of genome doubling as a factor generating cryptic diversity is just one reason why many plant biologists consider the estimate of 350,000 species of flowering plants to be a gross underestimate.

A similar example of cryptic species occurs in frogs (Fig. 4.2). Two frogs of the genus *Hyla* have nearly identical appearance, but differ in vocalization or call and also in chromosome number. The two frogs do not interbreed and are considered two species, *Hyla chrysoscelis* and *Hyla versicolor* (Gerhardt, 2005). Cryptic species are now thought to be common in frogs, and most examples do not involve changes in the chromosome number (D. Blackburn unpublished). For example, a number of cryptic species were detected and named in the African clawed frogs (*Xenopus*) (Evans et al., 2015); although similar in appearance, they differ genetically, in their calls, and in some cases even in the parasites that live on them. The prevalence of cryptic species has now been recognized throughout amphibians—they now are considered widespread in that group. Amphibians are therefore another example in a growing list of lineages in which cryptic species are a major source of unnamed and unrecognized biodiversity.

Cryptic species are often discovered quite by chance via DNA sequencing. For example, imagine that a DNA sequence of an individual

(A)   (B)

Figure 4.2 Cryptic species in frogs. Although similar in appearance, the two species differ in mating call and thus remain reproductively isolated: (A) *Hyla chrysoscelis* and (B) *Hyla versicolor. Wikipedia Free Commons.*

in a museum collection reveals that the specimen sequenced is very different from other collections of what has been named the same species (see also Chapter 5). An example in butterflies is shown in Fig. 4.3 (Burns et al., 2008)—four cryptic species were discovered using DNA sequence data. The species are clearly distinct in a phylogeny based on DNA data, but they appear identical to our eye. Similarly, other workers have discovered cryptic butterflies using DNA approaches (Hausemann et al., 2011). Cryptic species are now thought to be common in butterflies.

Nonspecialists often think that there are so many species of life still unnamed because of the huge amounts of unsampled biodiversity in the tropics, or in the deep sea, or in other remote areas of the world. While these under-collected geographic areas and hard-to-get-to places of the world are certainly an important factor in our lack of knowledge of species diversity on Earth, many new species of organisms exist under our noses as cryptic species in well-studied parts of the world. Cryptic species are all around us. We have only discovered the tip of the cryptic species iceberg.

WE ARE LIVING IN A BACTERIAL WORLD

Since the 1970s, biologists have recognized three domains or major branches of life—eukaryotes (described earlier), and two microbial lineages, bacteria and archaea (Fig. 4.4) (Woese and Fox, 1977). However, recent research has further reshaped this picture. Increased species sampling and more DNA data reveal that eukaryotes (our lineage) may have actually originated from within the archaea (Fig. 4.5). In fact, the eukaryotes are closely related to a group of archaea called the

Figure 4.3 Cryptic species of butterflies. Four cryptic species of butterflies in the genus *Perichares*. Males (columns 1 and 3) and females (columns 2 and 4), dorsal (left) and ventral (right) view. (1−4) *P. adela*; (5−8) *P. poaceaphaga*; (9−12) *P. geonomaphaga*; (13−16) *P. prestoeaphaga*. These species are all distinct genetically and are separate lineages of life, but look very similar in appearance. *From Burns, J.M., Janzen, D.H., Hajibabaej, M., Hallwachs, W., Hebert, P., 2008. DNA barcodes and cryptic species of skipper butterflies in the genus* Perichares *in Area de Conservación Guanacaste, Costa Rica. Proc. Natl. Acad. Sci. U.S.A. 105, 6350−6355, Fig. 3, doi: 10.1073/pnas.0712181105.*

Heimdallarcheota (a big name for small organisms). This recent work suggests, therefore, that there may be two major domains of life, not three—bacteria and archaea, with eukaryotes having originated from within the latter (Spang and Ettema, 2016). In other words, we are descended from archaea (Fig. 4.5).

By far most of life on Earth is microbial—although this enormous diversity is often unappreciated even by biologists. There may be 100 million or more archaea and bacteria, although only a small fraction of these have been discovered, and even fewer actually named. In fact, a huge problem for this part of the Tree of Life is that few species of the microbial world are named in the standard way that scientists have long named eukaryotic life since the time of Carolus Linnaeus, with Latin binomials. Binomial simply means "two names"—genus and specific epithet. *Homo*

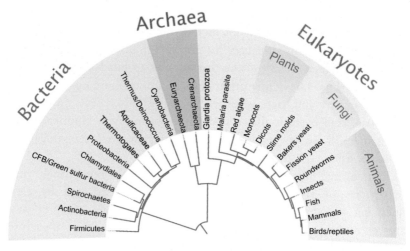

Figure 4.4 Traditional "three kingdom" summary Tree of Life based on ribosomal DNA sequence data showing three major groups (clades) of life. *Wikipedia Free Commons.*

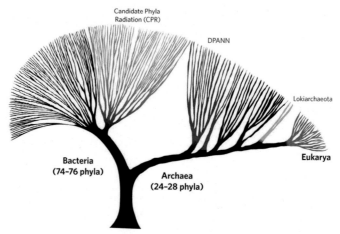

Figure 4.5 A possible new view of the Tree of Life with two major groups. In this tree, eukaryotes, labelled Eukarya in green (the lineage to which humans belong), are part of the Archaea group. That is, eukaryotes may be derived from an archaeal ancestor. *From Spang, A., Ettema, T.J.G., 2016. Microbial diversity: the tree of life comes of age. Nat. Microbiol. 1, 1–2, Fig. 1, doi: 10.1038/nmicrobiol.2016.56.*

sapiens is the name for our species, for example, with *Homo*, the name of the genus and *sapiens*, the specific epithet. Most archaea and bacteria are never given binomial or scientific names—they typically are only ever given a strain name/number (VPI 6807B; STJ. EPI. H7). As a result, scientists (who are often biased in thinking toward eukaryotes with their standard two-part names), do not include archaea and bacteria strains in the Tree of Life. This omission has important implications. When we consider a simple summary of the first attempt to build a comprehensive Tree of Life (Fig. 4.6), it appears as if there are far more eukaryotes than species of archaea and bacteria (in red). But this only reflects the relatively few archaea and bacteria that actually have been given binomial (scientific) names—those given binomials are generally well-known bacteria, such as *Escherichia coli*.

Undiscovered microbial species are everywhere. Interestingly, most new species of archaea and bacteria are now discovered by DNA sequencing of what are termed environmental samples. Soil or other materials are home to many archaea and bacteria. Sequencing of such soil samples reveals new microbial life identified only by a distinctive DNA sequence which is then simply given a strain number.

Furthermore, the human gut is host to large numbers of bacteria and archaea species (Eckburg et al., 2005; Backhed et al., 2005), now referred to as your microbiome. There are over 1000 common bacterial species in the human gut alone (Qin et al., 2010). Gill et al. (2006) in referring to the bacteria and archaea that we host estimated that the human intestine "contains $10^{13}-10^{14}$ microorganisms whose collective genome ('microbiome') contains at least 100 times as many genes as our own genome"; they further state that "humans are superorganisms whose metabolism represents an amalgamation of microbial and human attributes."

With all of these findings in mind, microbial diversity clearly dominates our world. Madonna, who famously sang, "we are living in a material world," was close to getting it right biologically—we are clearly living in a bacterial world. One of the most fascinating recent discoveries is that new bacteria and archaea are consistently discovered in all parts of the Earth—the surface of our planet, as well as the deepest parts of the ocean floor; these organisms even permeate into the Earth's crust.

HAVES AND HAVE NOTS: RESOURCES FOR MAJOR GROUPS OF LIFE

Our knowledge of the Tree of Life is far from evenly distributed—a few parts of the tree are known fairly well, but large parts of the tree are poorly understood by any metric. We could represent this with numbers

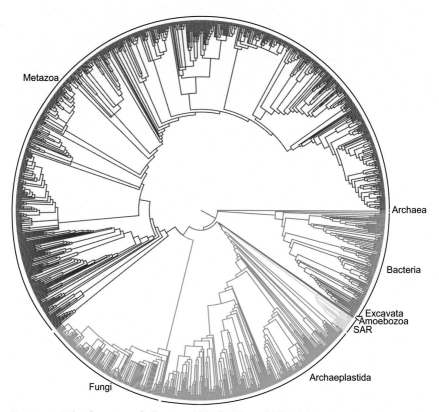

Metazoa

Archaea

Bacteria

Excavata
Amoebozoa
SAR

Archaeplastida

Fungi

Figure 4.6 The first tree of all named life from Hinchliff et al. (2015) (see Fig. 3.9). In this version, the major groups of life are shown using different colors. Bacteria (red) and Archaea (dark green) actually comprise many more species than indicated here. Most members of these groups are not formally given scientific names and hence were not included in the tree of all named life (see text). The remainder of the tree consists of different lineages of eukaryotes. Metazoa (gray-green) are a major group consisting of all multicellular animals with differentiated tissues. Fungi (blue) are another major group and are most closely related to Metazoa. Archaeplastida (light green) are another major group, consisting of traditional "plants"—plus the red algae, the green algae, and a small group of freshwater unicellular algae called glaucophytes. SAR (yellow), Amoebozoa (amoeboid protists), and Excavata (both dark green) are each relatively small groups of life. SAR is an abbreviation for a clade that includes stramenopiles (heterokonts--brown algae--are in this group), alveolates, and Rhizaria. Excavata includes free-living forms (*Euglena*) as well as some human parasites (*Giardia*). *Hinchliff, C.E., Smith, S.A., Allman J.F., Burleigh, J.G., Chaudhary R., Coghill L.M., et al., 2015. Synthesis of phylogeny and taxonomy into a comprehensive tree of life. Proc. Natl. Acad. Sci. 112, 12764–12769. Tree version provided by Stephen Smith. Modified from Hinchliff et al. (2015).*

to illustrate the point. But, perhaps a better way to illustrate how poorly understood parts of the Tree of Life are is to provide some examples.

A prominent bird biologist proudly proclaimed that among all organisms, birds have a wealth of resources, and they are very well-studied and well-known. This is indeed the case—birds have been the focus of considerable investment and study. No group of life is better known than birds. However, birds are a small part of the overall Tree of Life with just 10,000 species. Most bird species have been discovered, although a few new species are described each year, often to considerable fanfare. Most species of birds also have DNA sequence data available. Furthermore, there is a wealth of genetic resources for birds because a number of bird species have had their genomes completely sequenced.

Birds are tetrapods (Tetrapoda), a vertebrate group that also includes amphibians, reptiles, and mammals. Initial trees of relationship with nearly complete species sampling for all of the known tetrapods, based primarily on DNA sequence markers, are nearing completion (VertLife, http://vertlife.org/). Another group of tetrapods, the mammals (a small group of ~5500 species), are well sampled. Most species of mammals have been named, most have some DNA sequence data, and there are many mammals with complete genome sequences. New mammals are discovered and named each year and, as with birds, these findings typically get a lot of publicity. This is not the case for other less charismatic organisms, a case made by E.O. Wilson and others. Reptiles and amphibians are not as well-studied as mammals and birds, but are still generally well-known. For reptiles and amphibians, well over half of the known species have DNA sequence data, and more species continue to be described.

For the rest of life—which is by far the vast majority of its species—the state of knowledge for many groups is abysmal. Even among vertebrates, fish stand out as a poorly understood group. Although ~35,000 species of fish have been named, there may be over 10,000 still unnamed. Considering the ray-finned fish (the majority of all fish), of the species that are named, only about one third have DNA sequence data (Rabosky et al., 2018). Several fish genomes have been completely sequenced, so there are some genetic resources, as well.

Many species remain undescribed, and most of those that have been named have no DNA data of any kind. Genetic resources are marginal to nonexistent. For example, the sponges are better known than most nonvertebrates, but they are still poorly known. About 9000 species of sponges have been described, but 3000—6000 additional species likely exist (Bergquist, 2001). Only about 9% of sponge species now have any available DNA data and are actually included in trees of relationship (R. Thacker, pers. comm.). Similarly, 46,000 species of marine molluscs (including clams, oysters, scallops, mussels, squid, and octopuses) are known, but the estimated number of species is as high as 150,000-200,000 (e.g., Chapman, 2009).

Green plants (including green algae, mosses, liverworts, ferns, and seed plants, i.e., gymnosperms and flowering plants) are often considered well-studied, even by fellow biologists. Green plants afford a good example of how even scientists sometimes think we know much more than we actually do. The green plants may comprise half a million species and a billion years of evolution. They are the major drivers of terrestrial and many aquatic eco-systems (freshwater and many marine systems). They are also of enormous economic importance. Yet, by all metrics, green plants are poorly sampled and poorly understood in terms of sequence data. Perhaps the only well-sampled lineage is the gymnosperms (conifers and relatives) because of their small size (1000 species). Flowering plants have been considered well-studied by some, but are actually very poorly known—perhaps 15% of flow-ering plants remain undiscovered (Joppa et al., 2011). Sampling of green plants for DNA is less than 30% for "green" algae and only $\sim 30\%$ across the enormity of the flowering plants ($\sim 350,000-400,000$ species).

Fungi, insects, and microbial life (bacteria and archaea) are huge, poorly understood parts of the Tree of Life. For fungi (mushrooms and molds), there are 135,000 named species, but perhaps 1.5 to roughly 5 million still undescribed (Blackwell, 2011). So perhaps 2%–10% of the fungal diversity has been described. Many new species of fungi are rou-tinely discovered each year from what are termed environmental samples, just as noted for microbial life (above). A small sample of soil or rotting wood is sequenced, and the snippets of DNA recovered reveal many new species that have never been seen or sequenced before—just as with bacteria and archaea, vast numbers of fungal species continue to be discovered in this manner.

Protists, those mostly single-celled eukaryotes that remain after plants, animals, and fungi are considered, may be even more poorly understood than the fungi (Pawlowski et al., 2012). Approximately 14,000 species of protists have been described, but the estimated number is much higher, perhaps 1.6 million (Adl et al., 2007), and DNA coverage remains sparse; as with fungi, environmental sampling is again revealing much additional diversity (Pawlowski et al., 2012).

The hexapods (meaning six legs) are the largest group of arthropods and include insects, as well as three much smaller groups of wingless organisms, all of which were once considered insects. Hexapods are a big group and include approximately 1 million named species, but the likely number of hexapod species is considered to be much higher; there may be 2.6–7.8 million species of insects; beetles alone represent 0.9–2.1 million of those species (Stork et al., 2015). Furthermore, the named hexapod species are also poorly sampled for DNA—only 200,000 named species (20%) have available DNA data—so as with other obscure clades of life, much of the hexapod tree is poorly known. There are bright spots in the hexapod Tree of Life—being charismatic matters. In this regard, the butterflies are an exception among insects and represent a region that is well-known. More than 60% of the ~18,000 species of butterflies have DNA sequence data.

As noted, the bacterial and archaeal part of the Tree of Life is so immense that these parts of the tree may still be the most poorly known.

CHALLENGES TO COMPLETING THE TREE OF LIFE

The number of new species described per year by experts remains fairly steady at 14,000. Although this seems like a lot, at that pace it will take perhaps 900 years or more to name just what we think is here on our planet. This is a point made by many naturalists (e.g., Wilson, 2016). Progress on some groups may be faster than others. For example, although the yearly increment of newly described species of molluscs is steadily increasing, at the current pace it will take 300 years to name the estimated ~150,000 species of marine molluscs awaiting description (G. Paulay, personal communication).

A frequent misconception, among the general public, as well as among many scientists, is that a new species is named rapidly following its discovery in the field. But, this is far from the case. Based on a survey of new species described and selected among diverse lineages of eukaryotes (vertebrates, arthropods, other invertebrates, plants, fungi, and protists), Fontaine et al. (2012) showed that it takes on average 21 years between the collection of the first specimen of a new species in the field and its formal description. Many new species sit in museums for decades before formal description and naming, reflecting the fact that biodiversity science is drastically underfunded and understaffed (Wheeler et al., 2004). There are just not enough specialists to do the job, and the number of such experts is declining rapidly. As just one example, *The Wall Street Journal* recently noted that the U.S. is "running short of botanists" and highlighted the negative consequences of that shortage (www.wsj.com/articles/rhododendron-hydrangea-america-doesnt-know-anymore-1534259849).

The estimate that it will take over 900 years to name the remaining species on our planet does not take into account the rapid pace of extinction (see Chapter 6), which, in an ironic twist of fate, is making the task easier. For example, the number of species of amphibians is rapidly decreasing, although many new species continue to be discovered (McCallum, 2007). Both the named, but poorly understood, as well as the never described will never be known to humankind.

What is the solution to this major biodiversity challenge? Large pieces of the unknown part of the Tree of Life could rapidly be filled in with modestly increased investment in DNA sequencing of existing samples. For example, much of the existing land plant species diversity is represented as dried and pressed collections (Fig. 4.7; a herbarium specimen) in buildings or collections referred to as herbaria; these contain millions of plant specimens worldwide. With modern molecular methods, DNA sequences can be obtained from many of these dried samples. The record age for DNA sequence data from such a specimen is over 200 years in age. The plant portion of the tree could be rapidly completed with modest funding. DNA can similarly be obtained from many insect collections. DNA from just a leg of an insect can help place that insect in the Tree of Life.

Unfortunately, many existing fish, amphibian, and invertebrate collections cannot be used in the same way because these samples are collected and preserved in alcohol which degrades the DNA, making it largely unusable for study. Thus, for much of the Tree of Life an investment in new collections is needed to improve our knowledge of the unknown and the unsequenced.

Another very important solution to the biodiversity crisis noted here is more involvement of the public—citizen science. With dwindling institutional support from academic institutions and low funding from national funding agencies, the future of inventorying the species diversity of the world will depend on a major involvement of citizen scientists. This is a partnership that is not only crucial for saving biodiversity, but one that specialists welcome and encourage.

Figure 4.7 Image of a herbarium specimen. *Image by Kathleen M. Davis of specimen from the University of Florida Herbarium, FLAS 230552, Florida Museum of Natural History. Permission granted by University of Florida Herbarium.*

REFERENCES

Adl, S.M., Simpson, A.G., Farmer, M.A., Andersen, R.A., Anderson, O.R., et al., 2007. The new higher level classification of eukaryotes with emphasis on the taxonomy of protists. J. Eukaryot. Microbiol. 57, 189–196.

Backhed, F., Ley, R.E., Sonnenburg, J.L., Peterson, D.A., Gordon, J.I., 2005. Host–bacterial mutualism in the human intestine. Science 307, 1915–1920.

Bergquist, P.R., 2001. Porifera (Sponges). Encyclopedia of Life Sciences. John Wiley & Sons, Ltd.

Blackwell, M., 2011. The Fungi: 1, 2, 3 ... 5.1 million species? Am. J. Bot. 98, 426–438.

Burns, J.M., Janzen, D.H., Hajibabaei, M., Hallwachs, W., Hebert, P., 2008. DNA barcodes and cryptic species of skipper butterflies in the genus Perichares in Area de Conservación Guanacaste, Costa Rica. Proc. Natl. Acad. Sci. U.S.A. 105, 6350–6355.

Chapman, A.D., 2009. Numbers of Living Species in Australia and the World, second ed. Australian Biological Resources Study, Canberra.

Costello, M.J., May, R.M., Stork, N.E., 2013. Can we name Earth's species before they go extinct? Science 339, 413–416.

Dawson, M.N., 2012. Species richness, habitable volume, and species densities in freshwater, the sea, and on land. Front. Biogeogr. 4, 105–116.

Eckburg, P.B., Bik, E.M., Bernstein, C.N., Purdom, E., Dethlefsen, L., Sargent, M., et al., 2005. Diversity of the human intestinal microbial flora. Science 308, 1635–1638.

Evans, B.J., Carter, T.F., Greenbaum, E., Gvozdik, V., Kelley, D.B., McLaughlin, P.J., et al., 2015. Genetics, morphology, advertisement calls, and historical records distinguish six new polyploid species of african clawed frog (*Xenopus*, Pipidae) from West and Central Africa. PLoS ONE 10 (12), e0142823. Available from: https://doi.org/10.1371/journal.pone.0142823.

Fontaine, B., Perrard, A., Bouchet, P., 2012. 21 years of shelf life between discovery and description of new species. Current Biol. 22, R943–R944.

Gill, S.R., Pop, M., DeBoy, R.T., Eckburg, P.B., Turnbaugh, P.J., Samuel, B.S., et al., 2006. Metagenomic analysis of the human distal gut microbiome. Science 312, 1355–1359.

Grosberg, R.K., Vermeij, G.J., Wainwright, P.C., 2012. Biodiversity in water and on land. Curr. Biol. 22, R900–R903.

Hausmann, A., Haszprunar, G., Hebert, P.D.N., 2011. DNA barcoding the geometrid Fauna of *Bavaria* (Lepidoptera): successes, surprises, and questions. PLoS ONE 6 (2), e17134. Available from: https://doi.org/10.1371/journal.pone.0017134.

Hinchliff, C.E., Smith, S.A., Allman, J.F., Burleigh, J.G., Chaudhary, R., Coghill, L.M., et al., 2015. Synthesis of phylogeny and taxonomy into a Comprehensive tree of life. Proc. National Acad. Sci. 112, 12764–12769.

Judd, W.S., Soltis, D.E., Soltis, P.S., 2007. *Tolmiea diplomenziesii*: a new species from the Pacific Northwest and the diploid sister taxon of the autotetraploid *T. menziesii* (Saxifragaceae). Brittonia. 59, 218–225.

McCallum, M.L., 2007. Amphibian decline or extinction? Current declines dwarf background extinction rate. J. Herpetol. 41, 483–491.

Mora, C., Tittensor, D.P., Adl, S., Simpson, A.G.B., Worm, B., 2011. How many species are there on Earth and in the Ocean? PLoS Biol. 9 (8), e1001127. Available from: https://doi.org/10.1371/journal.pbio.1001127.

Pawlowski, J., Audic, S., Adl, S., Bass, D., Belbahri, L., Berney, C., et al., 2012. CBOL Protist Working Group: barcoding eukaryotic richness beyond the animal, plant, and fungal kingdoms. PLoS Biol. 10 (11), e1001419. Available from: https://doi.org/10.1371/journal.pbio.1001419.

Pérez-Ponce de León, G., Poulin, R., 2016. Taxonomic distribution of cryptic diversity among metazoans: not so homogeneous after all. Biol. Lett. 12, 20160371.

Qin, J., Li, R., Raes, J., Arumugam, M., Burgdorf, K.S., Manichanh, C., et al., 2010. A human gut microbial gene catalogue established by metagenomic sequencing. Nature 464, 59–65.

Rabosky, D.L., Santini, F., Eastman, J., Smith, S.A., Sidlauskas, B., Chang, J., et al., 2013. Rates of speciation and morphological evolution are correlated across the largest vertebrate radiation. Nat. Commun. 4, 1958. Available from: https://doi.org/10.1038/ncomms2958.

Sangster, G., Luksenburg, J.A., 2015. Declining rates of species described per taxonomist: slowdown of progress or a side-effect of improved quality in taxonomy?. Syst. Biol. 64, 144–151.

Scheffers, B.R., Joppa, L.N., Pimm, S.L., Laurance, W.F., 2012. What we know and don't know about Earth's missing biodiversity. Trends Ecol. Evol. 27, 501–510.

Soltis, D.E., 1984. Autopolyploidy in *Tolmiea menziesii* (Saxifragaceae). Am. J. Bot. 71, 1171–1174.

Soltis, D.E., Soltis, P.S., 1989. Genetic consequences of autopolyploidy in *Tolmiea menziesii* (Saxifragaceae). Evolution 43, 586–594.

Soltis, D. E., Soltis, P. S., Schemske, D., Hancock, J., Thompson, J., Husband, B., et al., 2007. Autopolyploidy in angiosperms: have we grossly underestimated the number of species? Taxon 56, 13–30.

Spang, A., Ettema, T.J.G., 2016. Microbial diversity the tree of life comes of age. Nat. Microbiol. 1, 16056. Available from: https://doi.org/10.1038/nmicrobiol.2016.56.

Stork, N., 1993. How many species are there? Biodiv. Conserv. 2, 215–232.

Stork, N.E., McBroom, J., Gely, C., Hamilton, A.J., 2015. New approaches narrow global species estimates for beetles, insects, and terrestrial arthropods. Proc. Natl. Acad. Sci. U.S.A. 112, 7519–7523.

Visger, C.J., Germain-Aubrey, C., Soltis, P.S., Soltis, D.E., 2017. Niche divergence in *Tolmiea* (Saxifragaceae): using environmental data to develop a testable hypothesis for a diploid-autotetraploid species pair. Amer. J. Bot.

Wheeler, Q.D., Raven, P.H., Wilson, E.O., 2004. Taxonomy: impediment or expedient? Science 303, 285.

Wilson, E.O., 2016. Half-Earth. Liveright Publ. Co, New York.

Woese, C.R., Fox, G.E., 1977. Phylogenetic structure of prokaryotic domain—primary kingdoms. Proc. Natl. Acad. Sci. U.S.A. 74, 5088–5090.

CHAPTER 5

The Value of the Tree of Life

"Nothing makes sense except in light of evolution"

Dobzhansky, 2013 [1973]

To this insightful phrase, numerous biologists studying biodiversity have added the corollary:

Everything in biology makes more sense in light of phylogeny.

As portions of the Tree of Life (e.g., vertebrates, butterflies) have been better resolved, together with the recent publication of the first rough draft tree of all named life (see Chapter 3), the scientific community, as well as the general public, have increasingly come to appreciate its value. Through yielding powerful insights into the past, the Tree of Life provides a means to interpret the patterns and processes of evolution, as well as the ability to predict the responses of life in the face of rapid environmental change. Broad knowledge of species relationships is fundamental, providing crucial new information regarding the discovery of medicines, combatting diseases, and crop improvement. This information has also had major impacts in the diverse fields of genomics, evolution, and development, while providing insights into the study of adaptation, speciation, community assembly, and ecosystem functioning. Given the many benefits, it is therefore hard to summarize, in a few words, the immense implications and applications of the Tree of Life to biology and human well-being.

All of these benefits of a better knowledge of phylogeny and the Tree of Life are made possible for the same reason that a clear understanding of your own family tree is important—knowledge of relationships matters. We all seem to have a fascination with family trees—who were my ancestors? How am I related to others? In addition, we all understand clearly that if a close relative has a disease that is inherited, say a certain cancer, then there is a good probability that we may have inherited the genes for that trait (Fig. 5.1).

In much the same way as understanding your family tree is enlightening, the Tree of Life similarly has informative and predictive value. We

The Great Tree of Life
DOI: https://doi.org/10.1016/B978-0-12-812553-3.00005-9

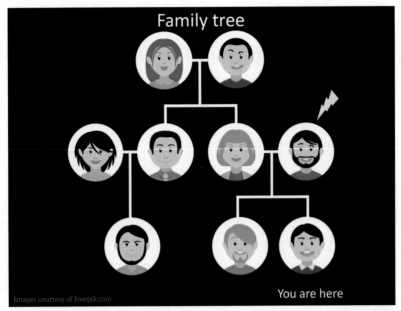

Figure 5.1 An imaginary family tree in humans showing that relationships matter for you as an individual. We all realize that if a relative or ancestor (lightning bolt) has a disease that is inherited genetically, there is a good likelihood that we inherited that trait. *Wikipedia Free Commons.*

can use the Tree of Life to inform, discover, and solve major problems that affect our own species. Closely related organisms may produce similar chemical compounds of medicinal value to our species. Close relatives of crops are the best source of genes for disease and drought resistance. Closely related strains of disease (e.g., the flu) will likely respond similarly to similar vaccines. And because closely related species will likely respond in similar ways to increases in temperature or drought stress, the Tree of Life can even be used to predict how organisms may respond to a rapidly changing climate. All of these examples and the predictive power of relationships depend on a firm knowledge of the Tree of Life.

Another way to look at the huge potential impact of the Tree of Life is that it represents the biodiversity equivalent of the human genome project. When the human genome project was initiated, there was considerable debate as to its actual value and whether it was worth the immense expense. To some skeptics, the human genome project represented a costly, lengthy, and basic research project with few practical outcomes. And while the cost of the first human genome was ~2.7 billion dollars and the project took over 10 years to complete (International Human

Genome Sequencing Consortium, 2004), the human genome project is now considered one of the great accomplishments of modern science. The human genome sequence has spurred on the discovery of the functions of numerous genes and the genetic underpinning of many diseases, transforming medical research. It has also impacted the study of human population genetics, revealing patterns of human migration through time. Furthermore, the human genome project also generated advances in DNA sequencing, making it now so routine that you can obtain your own genome sequence for less than $100. As sequencing costs decrease, your own genome sequence will soon be a fundamental part of your medical record. Efforts, in fact, are underway in China to sequence the genome of every person on Earth, and scientists have set goals of sequencing 10,000 plant species (the 10KP project; http://www.sciencemag.org/news/2017/07/plant-scientists-plan-massive-effort-sequence-10000-genomes), 10,000 vertebrate species (the G10K project; https://genome10k.soe.ucsc.edu/), and even all of life (the Earth BioGenome Project; https://www.earth-biogenome.org/).

Just as the sequencing of the entire human genome provided numerous and largely unanticipated new biological discoveries, reconstructing the entire Tree of Life has already fueled and will continue to fuel fundamental research and the development of practical tools to sustain biological diversity and enhance the quality of human life, whether through combatting disease, improving crops, or discovering new medicines. The multifaceted value of biodiversity to human well-being in this broad sense is crucial (see Grifo and Rosenthal, 1997; Chivian and Bernstein, 2008; Sala et al., 2012).

MEDICINES

There are multiple considerations regarding the importance of the Tree of Life and drug discovery. In fact, most of our medicines trace originally to chemical compounds from plants—humans did not invent these compounds, nature did. Species in nature have evolved numerous chemicals with diverse purposes and roles—the benefits range from defense to capturing prey. Those same chemicals may also be of value to our own species. Many of these useful compounds are discovered by chance through the basic research of an observant/inquisitive biologist, or traced to a long history of use in traditional medicines. On the one hand, it is therefore important to preserve the Tree of Life simply because there is so much hidden, as-of-yet untapped medical value in the species that compose the Tree of Life. Numerous chemical compounds of direct medical value to our own species await discovery.

Plants, fungi, and animals have been used for medicinal purposes for thousands of years. Indeed, modern research sometimes entails the detailed study of traditionally used medicinal plants (e.g., Zhang, 2002; Patwardhan and Mashelkar, 2009). However, as species are lost to extinction, that potential utility is also lost forever—hence, the importance of protecting the species that constitute the Tree of Life, simply for utilitarian reasons. Imagine the loss of a species that held the cure to a disease that might have saved your life, or the lives of your friends, relatives, or children. But imagine that this same species was driven to extinction before that knowledge was realized. How many potential medicines have already been lost due to species loss in the Anthropocene, and how many more will soon follow? If half of all plants (the major source of medicines) go extinct by the end of the century as predicted (see Chapter 6), imagine the human impact on medical loss alone!

As noted, new medicines often come from unexpected sources, and these make interesting stories in natural history and for the importance of protecting biodiversity. Consider a few examples provided below... and then contemplate what would have happened had these organisms gone extinct before this medicinal potential was even understood.

Hundreds of potential new compounds are discovered each year (Proksch et al., 2002), and excellent examples of the discovery of unanticipated new medicines trace to pitcher plants in the genera *Nepenthes* (Fig. 5.2A) (Eilenberg et al., 2006) and *Sarracenia* (Harris et al. 2012). Pitcher plants are carnivorous, widely known for trapping insects and other prey items and digesting them as a source of nitrogen. Prey items fall into a pool of fluid located inside the modified leaf (pitcher) (Fig. 5.2B) that contains digestive enzymes. Basic research has revealed that the pitchers of these plants have also evolved compounds that are antifungal—special enzymes that dissolve the cell walls of fungi. By producing these enzymes, the plants are able to inhibit fungal growth and by so doing the plants do not lose the resources in the prey trapped in the pitcher to fungi (Eilenberg et al., 2006). Significantly, these enzymes have shown great promise as new antifungal drugs in treating infections in our own species. Drugs derived from pitcher plants have been used to treat sciatic pain, symptoms of the herpes simplex virus (Mishra et al., 2013), diabetes (Muhammad et al., 2012), and tumors.

Figure 5.2 (A) Photograph of a species of *Nepenthes*, one of the pitcher plants, a plant with antifungal properties and medicinal value that were recently ascertained by scientists. (B) Photograph of a species of *Sarracenia*, another pitcher plant (distantly related to the *Nepenthes* in (A)); this is the source of sarapin, a compound of medicinal value. The close relatives of this species produce similar active compounds. (C) Photograph of *Conus magus*, the cone snail; neurotoxins that are naturally produced by this snail for paralyzing and capturing prey have been found to have uses in medicine as pain killers. *Wikipedia Free Commons.*

But the chemical constituents in these plants had until recently not been studied in great detail across all species of another group of pitcher plants, *Sarracenia* and related genera. A recent in-depth survey of numerous species and multiple genera of Sarraceniaceae pitcher plants not only provided a detailed survey of the chemical composition of numerous species, but also showed that chemical composition was highly correlated with phylogeny—that relationships are predictive (Hotti et al., 2017). This study represents a focused example of how phylogenetic trees are already forming the underpinning for more and more investigations of plants and medicinal compounds.

Another wonderful example of the value of unanticipated new medicines, as well as the importance of the basic research that yields these discoveries, traces to marine sea snails. Who would think that the venom of a poisonous sea snail, *Conus magus* (Fig. 5.2C) (McIntosh et al., 1982; Skov et al., 2007), could yield a new drug? A young scientist (B. Olivera) was fascinated by the ability of these organisms to produce a venom that can paralyze and kill prey. Years of research on the toxic compounds in these venoms resulted in the discovery of ziconotide, a nonaddictive drug, more powerful than morphine, that is used as a treatment for the chronic pain associated with cancer and AIDS.

There are similar examples of drug discovery in other marine organisms—for example, a bacterium that lives in close association with the bryozoan (*Bugula*) and secretes a substance that covers the larvae of the bryozoan and makes them distasteful to predators (Proksch et al., 2002) is the source of a potential Alzheimer's disease and cancer drug (Singh et al., 2008; Ruan and Zhu, 2012). As with many newly discovered chemical compounds useful to humans, the substance was discovered by basic, discovery-driven research as part of a survey of marine organisms for potentially beneficial chemical compounds (see https://pubs.acs.org/cen/coverstory/89/8943cover.html). Use of the Tree of Life and careful examination of close relatives of this bryozoan (or more specifically, close relatives of the bacterium) may be a useful way to find additional effective drugs.

Because closely related species often produce similar chemicals, the Tree of Life can be a road map to the discovery of new medicines. There are many such examples, but a classic case involves the Pacific yew (*Taxus brevifolia*; Fig. 5.3A) from western North America. This relative of pine trees is the original source of the drug paclitaxel (PTX0, sold using the brand name Taxol), a medicine that has been used to treat several types of cancer (Jordan and Wilson, 2004). From 1967 to 1993, nearly all paclitaxel was derived from bark obtained from the Pacific yew. But because the Pacific yew is uncommon and the process of obtaining the bark for medicine is a destructive process that kills the tree, the species is a problematic long-term source of the drug. Consequently, another source of

(A)   (B)

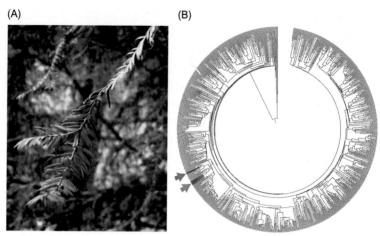

Figure 5.3 (A) The Pacific Yew, *Taxus brevifolia*, the original source of the cancer-treating drug Taxol. This species is rare, however. Using a tree of relationships, a close relative of this species from Europe (a species that is common) became the commercial source of Taxol until the drug was ultimately produced synthetically. (B) Chemical compounds of possible medical use to humans are often localized or clustered on the plant Tree of Life—these are "medicinal hotspots." Two such hotspots are shown here (in blue and green) in this circle tree of relationships of the flowering plants. Tree modified here to show useful chemistry of Solanaceae (tomato family) and Apocynaceae (dogbane or milkweed family). *(A) Wikipedia Free Commons. (B) Tree from Magallon, S., Gomez, S., Sánchez Reyes L.L., Hernández-Hern Andez, T., 2015. A metacalibrated time-tree documents the early rise of flowering plant phylogenetic diversity. New Phytol. 207. doi:10.1111/nph.13264. The tree is from Dryad and has no use restrictions. Tree modified here to show useful chemistry of* Solanaceae *(tomato family) and* Apocynaceae *(dogbane or milkweed family).*

the drug was needed. To find chemicals similar to what the Pacific yew produces, where do we look? Rather than randomly examining all 350,000 species of seed plants, the best approach is to use the Tree of Life to focus on the closest relatives of this species—and that is what was done. A commonly grown related species, the European yew, *Taxus baccata*, was then instead widely used in medicine until Taxol ultimately was made synthetically.

We also know that there are hotspots in the Tree of Life in which similar pharmacologically active compounds are produced by closely related species. These are excellent examples of how relationships matter. Consider the flowering plant families Solanaceae and Apocynaceae (indicated with blue and green in Fig. 5.3B); both are characterized by the production of diverse alkaloids that serve a defensive (antiherbivory) role.

In Solanaceae, the nightshade family (also the potato and tomato family), approximately 20 of the 90 genera have at least one or more species with medicinal properties. There are many well-known medicinal plants in the family including the nightshades (*Solanum*), belladonna (*Atropa*), pepper (*Capsicum*), jimsonweed (*Datura*), tobacco (*Nicotiana*), and mandrake (*Mandragora*); there is much medicinal potential remaining in the family with many species and genera still requiring detailed chemical characterization (Shah et al., 2013).

Just as is the case with the nightshade family (Solanaceae), the flowering plant family Apocynaceae (milkweed family) also has numerous plants with chemically active compounds. One common name for the family is "dogbane," referring to the fact that species in the family are poisonous for dogs. In fact, many members of Apocynaceae are poisonous, and again many have useful medicinal properties (Anderson, 1967). The family is a hotspot for chemical compounds, including cardiac glycosides used in heart ailments (e.g., the genera *Acokanthera*, *Apocynum*, *Cerba*, *Nerium*, *Thevetia*, and *Strophanthus*). Other genera contain active alkaloids that have applications in the treatment of cancer (*Catharanthus*), and still others have uses in relieving high blood pressure (*Rauvolfia*) or produce alkaloids with potentially beneficial psychoactive properties (*Tabernanthe*).

Despite the wealth of chemically active compounds in both Solanaceae and Apocynaceae, more medicinal discoveries are likely. For example, the applications in traditional medicine remain greatly underappreciated and understudied in species of both of these families, as well as in most families of plants in general. Just recently, in the Rajshahi district of Bangladesh, 14 species in 12 genera of Apocynaceae were identified with local medicinal uses (Rahman and Akter, 2015).

Fungi represent one of the great untapped parts of the Tree of Life for compounds of medicinal and other value to humans. Certain fungi have been major sources of useful medicinal compounds for human well-being (e.g., antibiotics) and offer enormous opportunity for discovering new compounds with diverse applications for human health (Katz and Baltz, 2016). To quote Tan et al. (2006), "Far from being mutually exclusive, biodiversity and genomics should be the driving force of drug discovery in the 21st century." We include the use of trees of relationship as another important component of the pathway to medicine discovery.

Rather than randomly surveying thousands of plants (there are roughly half a million green plants) or fungi (more than 120,000 species have been named, but there may be over 5 million species of fungi) for chemicals—the traditional approach—one can target hotspot areas or species closely related to known species of medical value. Such focused studies of known hotspots, coupled with the rapid evaluation of poorly known (dark areas) of the Tree of Life known to house useful compounds, may be the best path forward to the discovery of useful medical compounds (e.g., Cragg and Newman, 2013; Katz and Baltz, 2016).

But the practical value of the Tree of Life in terms of materials that may benefit our species includes far more than chemicals that provide medicines. The silk of spiders is incredibly strong and therefore has long been the source of interest for human needs in everything from lightweight shoes to support structures. Recent efforts that seek to combine knowledge of the structure of spider silk obtained from basic research with molecular genetic methods to produce these silk compounds in bacteria show signs of promise (Pennisi, 2017; Service, 2017).

DISEASE

Trees of relationships are now part of the first line of defense in combatting diseases. For example, when a new flu strain is detected, one of the first steps performed by researchers is to sequence the DNA of that strain and then gene sequences from that strain are compared using a phylogeny to other known viruses—in this way, a better understanding of the relationships of that strain can be quickly understood (see specific example below). That information allows more rapid vaccine development, based on knowledge of what has been employed as successful vaccines in what are determined to be closely related strains.

The appearance of new pathogens in humans following transfer to our species from other species poses special problems for vaccine development because the source of the new pathogen is not always clear. By sequencing these pathogens and using phylogenetic methods (tree building), the pathogen can be placed in the Tree of Life and its likely original host species determined. Consider SARS (severe acute respiratory syndrome), one of the classic examples of the use of phylogeny in tracing the origin of an infectious disease in humans. The presence of SARS in humans was first noted in China in November, 2002, and the disease ultimately spread around the globe to 30 countries and infected thousands of people worldwide. SARS resulted in hundreds of human deaths, as well as a worldwide health scare. It was clear that the original source was an animal species with transfer to humans, but the animal species that was the source of the

Phylogeny of SARS virus strains

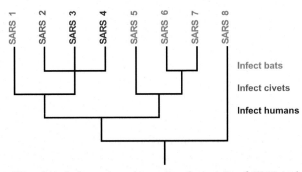

Figure 5.4 Use of a phylogeny to determine the origin of SARS in humans. Based on this simple tree, the closest relative of the strain in humans traces to two other mammals—civets or possibly bats. *Redrawn based on Eickmann, M., Becker, S., Klenk, H.D., Doerr, H.W., Stadler, K., Censini, S., Guidotti, S., et al., 2003. Phylogeny of the SARS coronavirus. Science 302, 1504−1505.*

virus that caused SARS in humans was initially unknown. However, phylogenetic analyses of DNA taken from viral strains occurring in various animals showed that human SARS traced its origin to civets and bats (Fig. 5.4; http://evolution.berkeley.edu/evolibrary/news/060101_batsars) (Eickmann et al., 2003; Guan et al., 2003).

Another classic example of the use of tree building in the study of disease is a case of HIV contracted in 1990 by a woman who had no real risk of obtaining HIV. This example is another "whodunit", solved using a phylogeny. The HIV virus evolves rapidly, and it is often possible to match the HIV in a patient to the original source, or human donor. It was determined using DNA sequence data and a phylogeny that the woman obtained the disease from her dentist, who was HIV positive. In fact, several other patients also obtained HIV from the same dentist, as clearly seen in a phylogenetic tree (Ou et al., 1992) (Fig. 5.5).

Phylogenetic analysis and the Tree of Life can also be used to combat influenza, a rapidly evolving virus. Because many strains of the flu virus have been stored over many years, DNA phylogenies can be produced to represent these known strains. As new flu strains emerge, they can be sequenced and then added to that phylogeny. In fact, via phylogenetic analyses, it has been possible to predict the likely dominant flu strain that will emerge the following year. This process of phylogenetic analysis is helpful in producing new flu vaccines and serves as critical information

(A)

(B)

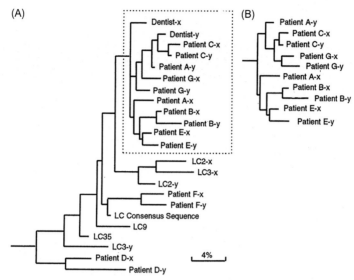

Figure 5.5 Use of DNA sequence data and a tree of relationships to track the transmission of HIV in a dental practice from dentist to some of the patients of that dentist. This is an example of the use of trees of relationship in detective work—to assess whether the dentist transmitted HIV to his patients. The tree indicates that he did. *Figure redrawn from Fig. 1 (A and B) of Ou, C.-Y., Ciesielski, C.A., Myers, G., Bandea, C.I., Luo, C.-C., Korber, B., Mullins, J.I., et al., 1992. Molecular epidemiology of HIV transmission in a dental practice. Science 256, 1165–1171.*

for designing the vaccine from one year to the next (Fig. 5.6; Fitch et al., 1997; Bush et al., 1999; Cui et al., 2016, www.cdc.gov/flu/professionals/laboratory/genetic-characterization.htm).

CONSERVATION

Phylogenetically Distinct Species

There are also multiple applications of the Tree of Life for conservation. The importance of phylogeny as a tool for conservation has been well reviewed elsewhere, including in other books devoted solely to that topic (e.g., Purvis et al., 2005, and the chapters therein). Given the large number of possible applications of the Tree of Life to conservation efforts, we will only focus on a few topics here.

Probably the most straightforward instances of the use of phylogeny in conservation are those examples involving the preservation of individual species. Take the white-winged warbler (*Xenoligea montana*), for example.

(A)   (B)

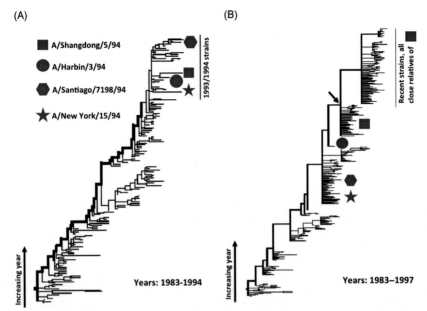

Figure 5.6 Predicting the evolution of human flu (influenza). Bush et al. (1999) examined the evolution of one domain of the human influenza or flu virus (domain HA1 of the H3 hemagglutinin gene). By building a phylogenetic tree, they found that a single dominant lineage persisted through time. Red symbols indicate several important recent strains. The tree in (A) shows the evolution of the HA1 domain from 1983 through 1994. Part (B) contains isolates from 1983 through 1997. In (B), the authors collapsed many branches of the tree for which they lacked strong support. Note that strain A/Shangdong/5/94 (indicated by a red square) descends from the node indicated by an arrow in tree B—this represents an uppermost node on (A) (A/Shangdong/5/94 is shown on both trees). This strain is further up the trunk of the tree in (B) than any of the other isolates from 1A—and it is also the isolate most closely related to future lineages (indicated by a vertical line). Thus, a tree can be used to predict future evolution of the virus. The authors also found evidence that 18 codons had been under selection in the past to change the amino acid they encoded. *Redrawn and modified from Bush, R.M., Bender, C.A., Subbarao, K., Cox, N.J., Fitch, W.M., 1999. Science 03 Dec 1999: 1921–1925 (see also Fitch, W.M., Bush, R.M., Bender, C.A., Cox, N.J., 1997. Long term trends in the evolution of H (3) HA1 human influenza type A. Proc. Natl. Acad. Sci. U.S.A. 94, 7712–7718).*

The white-winged warbler was long considered just one of many species of warbler ... however, use of DNA data and the Tree of Life showed it was not a true warbler (Fig. 5.7) (Klein et al., 2004). In a phylogeny of birds, the white-winged warbler occurred on its own branch, outside of the group or clade of true warblers. Based on these data, the bird was placed in its very own family. And because it only occurs on Hispaniola,

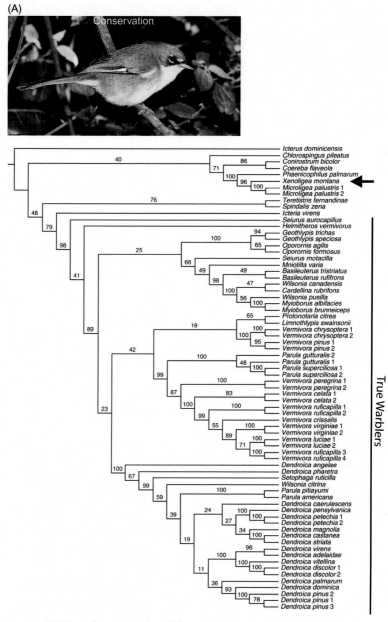

Figure 5.7 (A) The white-winged warbler, a bird that DNA data and the Tree of Life showed was not a true warbler. (B) Phylogenetic tree showing that the white-winged warbler (indicated by arrow) is not in the group (clade) with other warblers (true warblers indicated with vertical line). *(A) From Dave Steadman. (B) From Fig. 1 of Klein, N.K., Burns, K.J., Hackett, S.J., Griffiths, C.S., 2004. Molecular phylogenetic relationships among the wood warblers (Parulidae) and historical biogeography in the Caribbean Basin. J. Caribbean Ornithol. 17, 3–17.*

this distinct line of birds now merits extra conservation concern (Fig. 5.7).

A comparable example involving plants is *Amborella trichopoda*, a flowering plant that, until recently, was so poorly known that it does not even have a common name (consequently, we will refer to it simply as *Amborella* here). *Amborella* is a shrub or small tree and comprises perhaps only 12 populations, all restricted to the island of New Caledonia, which is over 1000 km off the east coast of Australia (reviewed in Soltis et al., 2008, 2017; *Amborella* Genome Project, 2013). *Amborella* remained largely unstudied and poorly understood until just a few decades ago, so its relationships to other flowering plants were unclear. Some researchers thought it was a member of the avocado or laurel family (Lauraceae), while other plant experts placed it in its own family (Amborellaceae), but still close to Lauraceae (reviewed in Soltis et al., 2008, 2017).

For decades, *Amborella* remained a poorly understood plant of very little broad interest to other scientists or to the public. But when DNA studies in the 1990s finally showed the placement of *Amborella* in the plant Tree of Life, that all changed. *Amborella* has the distinctive position as sister to all other living flowering plants (Fig. 5.8). That is, *Amborella* is to flowering plants what the duckbilled platypus is to mammals (Warren et al., 2008). This unique position in the flowering plant Tree of Life (Fig. 5.8) heightened conservation concern of *Amborella*, as this single surviving lineage of early angiosperms can provide critical insights into flowering plants.

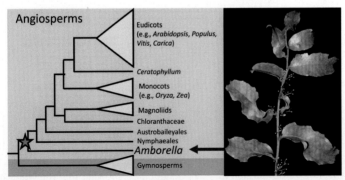

Figure 5.8 The flowering plant *Amborella trichopoda* and its pivotal position in the flowering plant Tree of Life as sister to all other living flowering plants. *Plant photograph courtesy of S. Kim, Sungshin University.*

Amborella has now been studied intensively because of its pivotal phylogenetic position. The complete genome of *Amborella* has been sequenced to provide an evolutionary reference for other flowering plant genomes and for applications in crop science (*Amborella* Genome Project, 2013). *Amborella* provides a baseline for comparison to help interpret gene and genome evolution in more derived flowering plants. Modern crops have highly complex and modified genomes—think of them as sophisticated fighter planes. If you know nothing about flight, these genomes are hard to interpret. *Amborella* is the genome equivalent of a biplane, providing the context needed to interpret the genomes of more complex, derived angiosperms (e.g., our crops) (*Amborella* Genome Project, 2013). All of this importance was only attached to *Amborella* once its position in the Tree of Life was ultimately realized.

Conservation Genetics—Breeding Programs

The Tree of Life can also play an essential role in designing breeding programs for the last survivors of a species. This lesson was learned too late for the dusky seaside sparrow, once native to the Atlantic Coast of Florida. When the numbers of this bird dwindled to a few males, a captive breeding program was designed using females from a geographically close subspecies of seaside sparrow, a program that failed (Avise and Nelson, 1989). Unfortunately, once a tree of relationships was later constructed for these birds, it revealed that the dusky seaside sparrow was most closely related to a seaside sparrow on the Gulf Coast of Florida rather than to geographically closer seaside sparrows as had been assumed. The captive breeding program should have involved this more closely related Gulf Coast subspecies instead, but insights from phylogeny came too late to save the dusky (Fig. 5.9).

There are other examples of how poor knowledge of organismal relationships can be deadly to those species. Conversely, other studies employing DNA markers and phylogeny have been useful in conservation, showing, as just one example, in freshwater mussels in Europe that populations actually remain of a rare species some had thought extinct; DNA data also revealed that those remaining populations are clearly genetically distinct from another closely related, more geographically widespread species. The two species had been confused because of their similar appearance (Prie et al., 2012).

Figure 5.9 Dusky seaside sparrow, an example of an organism that is now extinct where conservation efforts were not guided by a tree of relationships. Breeding programs to save the sparrow unfortunately did not pick the closest living relative of the dusky. *Free image online.*

Discovering Cryptic Species

For centuries, humans have identified and distinguished species based on physical appearance (morphology). But, two distinct species can appear so similar that biologists may have mistakenly considered them a single species. This topic was covered in Chapter 4. But to review, there are frog species that look nearly identical, but differ in their vocalizations, or calls; although difficult for us to distinguish by appearance, the frogs have no problem distinguishing each other. Similarly, there are plant populations that are very much alike in appearance and differ only in chromosome number—they cannot interbreed and have different environmental ranges. Distinct butterfly species may differ only slightly in spot pattern, or species that appear similar may have different behavior patterns. There are many other examples of cryptic species in other lineages, and they abound in nature, as reviewed in Chapter 4. As a result, we have grossly underestimated the number of species on our planet, even in those parts of the Tree of Life where we think we know a great deal (see Chapter 4). By using DNA sequence data and the Tree of Life, it is possible to more rapidly identify cryptic species. DNA sequence data and tree building (often referred to as phylogenetic reconstruction) have been applied to address this question of cryptic species. This approach may reveal that two entities that look similar differ in DNA sequence and have different placements in the Tree of Life. This type of investigation has been applied to several geographic regions thought to harbor large amounts of still undetected biodiversity; using DNA data and comparisons with species already in the Tree of Life can identify such cryptic biological entities quickly (e.g., Lahaye et al., 2008). Such DNA approaches and use of the Tree of Life are becoming more commonly applied by specialists in the continuing search for as yet undiscovered and unnamed species.

Protecting Areas Rich in Biodiversity

Another use of the Tree of Life is in assessing biodiversity and determining which areas of the Earth are most important to protect. Biologists already have a general understanding of where many biodiversity hotspots are located, although new ones are often discovered and proposed (e.g., the North American Coastal Plain). Compilations of massive amounts of natural history specimen data can help with these assessments of where there are major concentrations of biodiversity (e.g., Ulloa et al., 2017; Givnish, 2017). Phylogeny is also critical in determining which regions are home to the biggest part of the Tree of Life (e.g., Vázquez and Gittleman, 1998; Allen et al., in press; Lu et al., 2018).

An essential new conservation goal is to determine how much of the Tree of Life is present in any given area. This can be calculated using a phylogenetic tree and a measure called phylogenetic diversity (PD) (Faith, 1992; Mishler et al., 2014). Exploring PD is now a major research theme in biodiversity studies. An estimate of PD is not the same as an assessment of either the total number of species or the number of rare species in an area, both commonly used metrics when discussing conversation. While all three measures are incredibly important, they reveal very different things. Imagine two areas in a very simple example (Fig. 5.10). Which do you protect? The most species, the rare species, or the largest swathes of the Tree of Life (i.e., protecting PD)? While protecting as many species as possible, especially rare species, is inarguably important, it may be more effective to spend resources to protect PD.

Forest et al. (2007) provide an early example of the use of PD and the Tree of Life for conservation. This study showed important implications for conservation in the Cape Region of South Africa, a floristically rich

Area 1: Oaks

Area 2: More Phylogenetic Diversity

Figure 5.10 Explaining phylogenetic diversity (PD). Consider two hypothetical geographic areas of the same size with the same number of species, all depicted here. The area at the top has many species of oaks, some of them very rare. The area shown on the bottom has a wide diversity of species. Although the area on the top may have more rare species, the area on the bottom has much higher coverage of the Tree of Life, and much higher PD.

area that is a well-known for its biodiversity. Using a phylogeny for the plants of the region, Forest et al. (2007) revealed that the area with the highest PD did not correspond to the area with the greatest number of rare species, further illustrating that while these measures of biodiversity are not correlated, they are both critical for conservation.

To provide additional examples of the study of PD, we use recent studies from Florida and China to illustrate use of the Tree of Life to assess where PD is distributed. China is amazing in its extent of plant diversity—the country is home to nearly 10% of the ~350,000 species of flowering plants in the world. Using a large phylogenetic tree for all the genera of China, as well as a second tree for the 26,978 named flowering plant species there, it was possible to assess how the major components of the present-day vegetation came together and what areas are home to the highest PD. In their study, Lu et al. (2018) discovered that the flora of eastern China harbors many of the older lineages in China; distant relatives often co-occur in this part of China, and it has higher PD than does western China. In contrast, western China shows the co-occurrence of closely related plants that appear to be the result of recent mountain uplift, the formation of the Qinghai-Tibetan Plateau, and a resulting rapid radiation in that area; as a result, it has lower PD than observed in eastern China (Fig. 5.11). These results are important because they

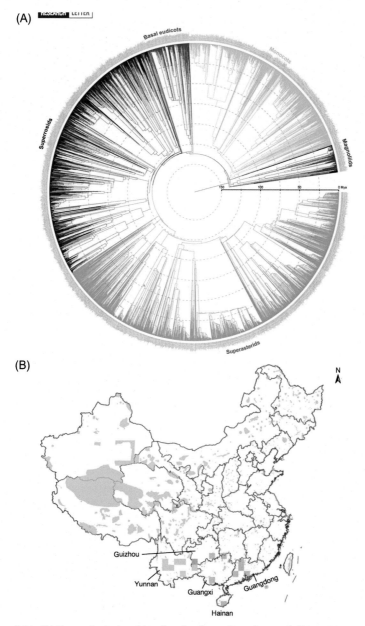

Figure 5.11 (A) Tree of relationships for the flowering plants of China. Major groups (clades), including magnoliids, monocots, superrosids, and superasterids, as well as the basal eudicot grade, are indicated with different colors. (B) The distribution of plant phylogenetic diversity (PD) in China. Grid cells with the top 5% highest phylogenetic diversity at the genus (pink) and species (blue) levels. Protected areas are highlighted in green, showing that areas of highest PD in China are not protected. *(A and B) From Lu, L.M., Mao, L., Yang, T., Ye, J.F., Liu, B., Li, H.L., e al., 2018. Evolutionary history of the angiosperm flora of China. Nature 554, 234–238. doi:10.1038/ nature25485. (B) Maps adapted from National Administration of Surveying, Mapping and Geoinformation of China (http://www.sbsm.gov.cn; review drawing numbers: GS (2016)1576, GS(2016)1549).*

identify areas of high PD and provide important data for conservation efforts. For example, the areas of high PD in western China are largely protected, but the same is not the case for areas of high PD in eastern China (Fig. 5.11). This lack of preservation of land in eastern China is often due to urbanization and the division of areas into different administrative units. Because the PD data indicate such high biodiversity in eastern China, an effective and supportive argument can be made for the necessity of more connectivity between national parks and nature reserves in that area (Lu et al., 2018).

Florida represents another good example of estimating PD for conservation purposes and future planning. With its many diverse habitats, Florida is also home to a highly diverse flora, including over 4300 species of vascular plants. Various locations in Florida are considered hotspots of biodiversity, including a portion of coastal Florida that is part of the North American Coastal Plain biodiversity hot spot (Noss et al., 2015) and the panhandle of Florida (Stein et al., 2000). This illustrates a problem with biodiversity hotspots—although important for conservation, there is no standard measure of assessment or calculation. This is why a measure such as PD is so valuable—it can be clearly defined, measured, and compared.

Importantly, Allen et al. (in press) found that PD was actually higher in the northern peninsula of Florida than in the coastal plain or peninsular hotspots (Fig. 5.12B). This can be explained in terms of the diverse habitats that come together in northern peninsular Florida (Allen et al., in press). In contrast, the Lake Wales Ridge, the high-elevation backbone of central Florida, is an area of low PD (see arrow on map of Florida in Fig. 5.12B). This ridge is an area well known for sand-scrub endemics— it is home to numerous closely related species, so the low PD is expected. Similarly, the Everglades at the southern tip of Florida also have low PD, reflecting its rather uniform habitat and species composition. The presence of a large number of closely related species in a region results in low PD, even if some of these species are rare (Fig. 5.12).

It is important to stress that no single biodiversity measure alone is right or wrong, whether it be the total number of species, the number of rare species, biodiversity hotspot designation, or estimate of PD; all are important ways to measure biodiversity. However, because PD is clearly defined, relies on the Tree of Life, and serves as a way to discover and hopefully preserve areas that are home to big pieces of the Tree of Life, it is important to measure PD and use it in making conservation decisions.

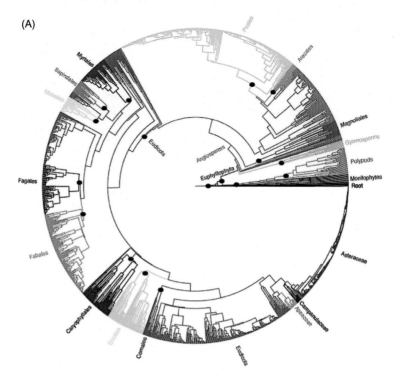

(A)

(B)

Phylogenetic Diversity (PD) in Florida

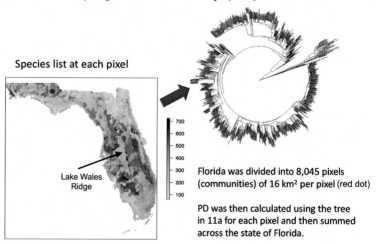

Species list at each pixel

Lake Wales
Ridge

700
600
500
400
300
200
100

Florida was divided into 8,045 pixels
(communities) of 16 km² per pixel (red dot)

PD was then calculated using the tree
in 11a for each pixel and then summed
across the state of Florida.

Figure 5.12 (A) Phylogeny of vascular plants in Florida—the Florida vascular plant
Tree of Life. This tree was constructed based on gene sequence data. It contains
1498 species (38% of the species of vascular plants in Florida); 685 genera (44% of
the total); and 185 families (78% of total). Major groups of plants are labeled and col-
orcoded. (B) Using the tree in 12A to show where plant phylogenetic diversity (PD) is
located in Florida. Florida was divided into 8045 pixel (communities) of 16 km² (the
red dot in northwest Florida shows one such pixel). A species list for each pixel was
then generated, and using the Florida Plant Tree of Life, the phylogenetic diversity
of each pixel (community) was calculated and then summed across the state of
Florida. If a species on the tree is in the pixel, it is colored red on the tree. Darker
green on the map represents higher PD. The arrow on the map indicates Lake Wales
Ridge, an area with low PD (light green). *(A and B) From Allen et al. (in press).*

At a time in which the leaves (species) of the Tree of Life are quickly disappearing or under threat (see Chapter 6), measuring PD across large regions of the globe is the best way to assess which regions to protect to save as much of the Tree of Life as possible.

RESPONSE TO CLIMATE CHANGE

Phylogenetic trees are now also essential tools in ecology. As closely related species often respond in similar ways to changes in the environment, scientists can employ phylogenetic relationships to predict how species may respond to such events as higher temperatures or less water. In a rapidly changing world, this important implementation of the Tree of Life cannot be understated.

A group of flowering plants called Saxifragales—or saxifrages, sometimes amusingly referred to as sexy-frages, due to the plants' attractiveness—serves as a prime example of how phylogenetic relationships in the Tree of Life can be useful in projections of response to climate change (Fig. 5.13). This small group of about 2500 flowering plant species contains some well-known woody plants, including sweet gum, currants, and witch hazel, as well as ornamentals including peonies, piggyback plant, coral bells, *Sedum* (stonecrops), and mother of thousands (Fig. 5.13). Despite the small number of species in this group, the saxifrages have enormous habitat variation; some are temperate forest trees, others are desert succulents, another group arctic alpines, and still others aquatic.

Using a phylogenetic tree for Saxifragales, it is clear that habitat shifts are very rare in these plants (Fig. 5.14). This habitat constancy is obvious just looking at the colors that correspond to habitat types in Fig. 5.14. The colors correspond closely to lineages within the saxifrages. Once a habitat shift was made over evolutionary time, for example, to a desert or aquatic or alpine habitat, lineages do not switch (or only rarely switch) out of that habitat—changes in habitat are canalizing events, and those habitats are the ones in which that lineage remains for millions of years.

We can illustrate the rarity of major niche shifts or changes in organisms by mapping the temperature preferences of species in the saxifrages on a phylogenetic tree that encompasses the ~110—120-million-year history of the group. This exercise is valuable because it shows that the group originated from a temperate ancestor with multiple changes or adaptations to cold temperatures and to warm temperatures (Fig. 5.15). But, once a change was made to cold temperatures, the plants do not

Figure 5.13 The flowering plant group Saxifragales; photographs of plants showing the tremendous diversity in a small group of approximately 2500 species. (A) *Heuchera* sp. ("*Heuchera* × *bryzoides*"), (B) *Tolmiea menziesii* Torr. and A. Gray, (C) *Kalanchoe blossfeldiana* Poelln., (D) *Sedum middendorffianum* Maxim., (E) *Liquidambar formosana* Hance, (F) *Ribes rubrum* L., (G) *Paeonia* "Red_Charm" (*Paeonia lactiflora* Pall. × *P. officinalis* L.), (H) *Hamamelis* × *intermedia* Rehder, (I) *Saxifraga caesia* L. Photographs from *Wikipedia Free Commons. Compiled by D. Soltis and E. Mavrodiev, Florida Museum of Natural History, University of Florida.*

quickly switch to growing in warm temperatures. This is also the case when a few lineages adapted to warm temperatures millions of years ago—they do not then switch to cold.

This work illustrates what biologists refer to as *phylogenetic constraints*. This seems at first glance an intimidating term, but it simply means that features that have already evolved in these organisms will play a major role in what those lineages can do in the future. That is, the evolutionary history of a group (or phylogeny), indeed the entire Tree of Life, may constrain future evolutionary options. In the case of saxifrages, once a lineage is cold-adapted, it is hard to switch—that is a phylogenetic constraint. These are the additional major features that a phylogeny can reveal. Similar findings have been reported for diverse geographic areas as well as other very different lineages of life, including rattlesnakes (e.g.,

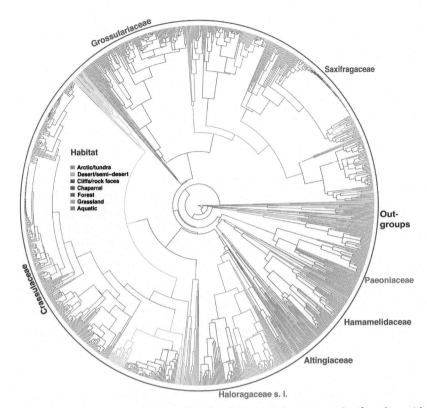

Figure 5.14 A tree of relationships for the flowering plant group Saxifragales, with habitat type mapped onto the tree using different colors. The group is highly diverse in habitat, with some species in desert, forest, arctic, and even aquatic habitats. Note that the colors correspond very closely to groups or clades (an ancestor and its descendants) in the tree. This shows that changes in habitat are rare, and when they do occur, they are often canalizing events. *For background see Soltis, D.E., Mort, M.E., Latvis, M., Mavrodiev, E.V., O'Meara, B.C., Soltis, P.S., Burleigh, J.G., and Rubio de Casas, R.R., 2013. Phylogenetic relationships and character evolution analysis of Saxifragales using a supermatrix approach. Am. J. Bot. 100, 916–929 and Rubio de Casas, R. R., M. E. Mort, and D. E. Soltis. 2016. The influence of habitat on the evolution of plants: a case study across Saxifragales. Annals of Botany 18(7):1317–1328.*

Kuntner et al., 2014; Lawing and Polly, 2011; Willis et al., 2008). These other studies, in addition to the case study of Saxifragales, indicate that it will be very difficult, and likely improbable, for many lineages of life to adapt to rapid climate change. The future of many lineages of life is indeed bleak—and using the Tree of Life, scientists can actually predict which lineages are most likely to have the greatest difficulty adapting to

Saxifragales: Ancestral Niche

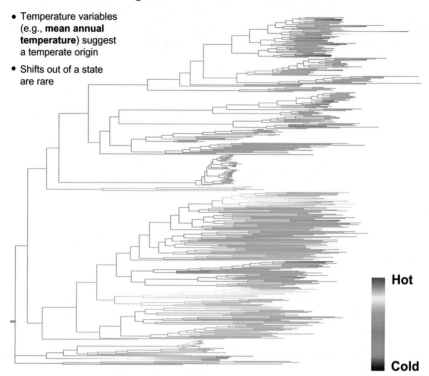

- Temperature variables (e.g., **mean annual temperature**) suggest a temperate origin
- Shifts out of a state are rare

Hot

Cold

Figure 5.15 The response of members of Saxifragales to temperature using a tree of relationships (see above) and plotting the average mean temperature of species occurrence on the tree (the tree is now horizontal in orientation, not circular as in Fig. 5.14). By using a tree of relationships, it is clear that the group originally evolved over 100 million years ago in cool temperatures (probably as forest trees). As the clade evolved and new species emerged, some lineages became adapted to very cool temperatures (blue) and others to warmer temperatures (yellow/red). But these evolutionary changes are canalizing events (as seen also in Fig. 5.14). Cool-adapted lineages have remained so for millions of years and do not spawn new species that are warm-adapted. The future for such lineages under scenarios of rapid climate change is bleak. *From Ryan Folk, Florida Museum of Natural History, University of Florida.*

any dramatic change in climatic factors, whether it be temperature or moisture.

CROP IMPROVEMENT

The Tree of Life is also important for crop improvement. If crop scientists want to find a way to make a crop more water-efficient—an important consideration today—or introduce disease resistance genes, where does one look for a source of genes (germplasm) for these goals? A common approach is to use knowledge of the Tree of Life to determine the crop's closest relatives to see if a closely related species has the trait of interest.

The critical importance of knowledge of phylogeny, the Tree of Life, to agriculture and crop improvement is vastly underappreciated. Some of our crops would not have survived without the introduction of genes from wild relatives. Consider sugarcane. "If no germplasm from wild relatives had been used there would probably not be a viable sugarcane industry in any place in the world" (J.D. Miller, USDA).

Also consider cultivated squash. Cultivated squash species require a considerable amount of water, but phylogeny reveals a group of close relatives of the cultivars that are drought-tolerant. These related, drought-tolerant species could provide a source of germplasm for crop improvement to help make the cultivated squashes more water-efficient (Fig. 5.16).

Figure 5.16 Cultivated species of squash and pumpkin (*Cucurbita*) require a great deal of water. Using a phylogeny for all species of *Cucurbita*, the cultivars were found to all be part of a mesic or wet-adapted clade. However, a tree of relationships reveals close relatives of the cultivars that are dry-adapted—these species are from the arid Southwest of the United States, as shown here. These closely related dry-adapted species are possible sources of germplasm for breeding purposes and ultimately the production of cultivars with greater water-use efficiency. This represents one example of how the Tree of Life can be used in agriculture. *From Heather Rose Kates, Florida Museum of Natural History, University of Florida.*

Legumes (Fabaceae; the bean family) represent the second most economically important family of plants after the grasses. Traditional efforts in crop breeding involved assessing genetic diversity among lines of domesticated species. Recently, however, the importance of phylogeny has increasingly emerged in legume breeding efforts (Smykal et al., 2015). Knowledge of the legume Tree of Life is crucial for understanding the origin, evolution, and ecology of legume crops. Many legume crops still have wild relatives in nature (unlike some of our cereal and mustard crops), so identifying these relatives can provide crucial information to improve disease resistance, water-use efficiency, and yield.

An excellent example of how the impact of knowledge of phylogeny can influence applied research can be seen in efforts to move the process of nitrogen fixation from legumes to other crops. Farmers and nonfarmers alike remain aware of the traditional process of crop rotation, in which a legume crop (e.g., alfalfa) is grown after a grain crop, such as corn, to enrich the soil with nitrogen. This is made possible because many legumes have special structures on their roots called nodules, which house special bacteria that can convert nitrogen from the air into nitrates that the plant can use. This adaptation enables legumes to thrive in poor soils. In many regions, crop rotation has largely been replaced by the application of large amounts of nitrogen-containing fertilizers.

Although these fertilizers have vastly increased the ease of growing the same crop annually, they also have shortcomings. The runoff containing these fertilizers, for instance, can prove immensely damaging to aquatic ecosystems. Another drawback of these fertilizers is their high cost in terms of both energy and finances. These issues have helped heighten a decades-long interest in understanding the mechanism used by legumes to produce nodules, moving this capability to crops that lack it, and then growing these plants without the use of fertilizers. Imagine the positive possibilities of growing numerous crops in poor or marginal soils at a potentially lower cost without the negative environmental effects of fertilizers.

It is also important to note that legumes are not the only plant family with species that have the ability to house nitrogen-fixing bacteria in root nodules. Nine other flowering plant families—including *Ceanothus* (wild lilac), members of the rose family, and a relative of cannabis—have similar capabilities. Because these plants look very different, they were long considered distantly related, implying that nodule production evolved over and over again. However, phylogenetic studies revealed—to the surprise of most botanists—that all of the plants that house nitrogen-fixing bacteria in root nodules are very closely related, meaning that the ability to produce root modules may have evolved just once. There is a single underlying mechanism that must be elucidated to move a cassette of genes from plants that can fix nitrogen via their bacterial partners to non-nitrogen fixers. In the past, research focused just on legumes to increase our understanding of nodulation. Now, however, research is focusing on these other members of the nitrogen fixing group to find the commonalities among species representing all ten families of plants with this ability. Importantly, this research strategy was made possible by improved understanding of the Tree of Life.

Trees of relationships are similarly used to determine the close relatives of plants of horticultural importance, and that information can then be used to select wild species and traits of interest for improvement of ornamental species (Handa et al., 2006; Takashi et al., 2006). For example, Japanese azaleas (*Rhododendron*; family Ericaceae) are widely cultivated and prized for their flowers. By using DNA data and building a tree of relationships, the closest relatives of the cultivated species were determined. These close wild relatives have desirable traits for breeding, including flowering time, different flower shapes and colors, and traits of cold and shade tolerance that could be used for improvement of Japanese azaleas.

PHYLOGENETIC DETECTIVE: FORENSICS AND THE TREE OF LIFE

The Tree of Life now plays a major role in detective work. In a manner similar to that of your favorite TV show involving forensics, DNA markers and the Tree of Life can be used to solve crimes. In what is essentially biodiversity forensics, a tool called DNA barcoding can be used to identify a species from a small amount of material or tissue. In this process, a genetic marker (a DNA sequence from one or more regions of the genome) is used to determine which species in the Tree of Life matches the DNA of the organism of interest (Fig. 5.17). Examples of the application of DNA and the Tree of Life in detective work range from the sale of mislabeled fish, to the smuggling of drugs, to the illegal harvesting of protected species. These approaches can also be used for conservation and the nondestructive assessment of the frequency of rare/endangered species. Several examples are reviewed below.

When you buy fish at the market or order it at a restaurant, how do you know it is cod, tuna, mackerel, or whatever species you think you

Figure 5.17 A diagrammatic representation of a DNA sequence barcode. In this example, the barcode sequence of each species shown (photographs) is the top sequence—that barcode sequence distinguishes it from other related species, listed below the top sequence.

are actually purchasing? DNA sequencing of tissue from the fish sample can be matched to known DNA sequences available for commonly consumed fish to determine which species on the Tree of Life you are truly getting and whether or not the fish has been legally obtained (Fig. 5.18; Stern et al., 2017). The results are alarming. A study by Oceana (see http://oceana.org/) revealed that the species of one third of the 1215 samples of seafood that they examined across 674 stores and restaurants taken from 21 states in the United States were incorrectly labeled (Warner et al., 2013). The results for sushi restaurants are particularly disturbing. Every sushi restaurant sampled from Chicago, New York, and Washington, DC, had at least one mislabeled fish (Stern et al., 2017; see http://www.smithsonianmag.com/science-nature/how-dna-testing-can-tell-you-what-type-of-fish-youre-really-eating-378207/). In Los Angeles sushi restaurants studied over a three-year period, researchers found that 47% of the fish samples were incorrectly labeled (Willette et al., 2017).

DNA sequence data and the Tree of Life can also be used for detective purposes pertaining to biological materials that are being brought through customs illegally. For instance, tobacco (*Nicotiana tabacum*) is of major economic importance worldwide; cigarettes alone are valued at more than 700 billion dollars a year globally. However, the smuggling of tobacco products across borders to avoid paying taxes has become an increasing problem. Government tax losses on a worldwide basis on tobacco products are estimated at $31 billion. Using DNA data and the Tree of Life, it is possible to identify plant fragments being transported across borders as tobacco (Biswas et al., 2016). This approach using DNA sequencing

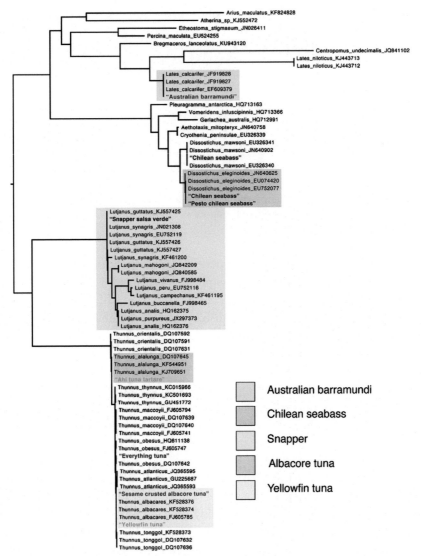

Figure 5.18 Tree of relationships based on DNA sequence data showing how the true identity of fish used in restaurants can be determined. DNA from a fish labeled "Snapper" in the market should appear in the tree shown here with other snapper samples. If that DNA appears with another species, the buyer has been misled—in fact, many of the fish on restaurant menus were found to be incorrectly named. Open access: *Stern, D.B., Nallar, E.C., Rathod, J., Crandall, K.A., 2017. DNA barcoding analysis of seafood accuracy in Washington, D.C. restaurants. Peer J. 5, e3234. https://doi.org/10.7717/peerj.3234.*

and the Tree of Life can also be used to detect the smuggling of other drugs, such as cocaine and fragments of marijuana (*Cannabis sativa*) (Coyle et al., 2003; Linacre and Thorpe, 1998; Staginnus et al., 2014).

Illegal logging and trading of timber are now major problems worldwide, resulting in threats or loss of rare plant species and threatened populations (UNOC Best Practice Guide for Forensic Timber Identification, 2016). The logging of dipterocarp forests in southeastern Asia serves as a prime example. Dipterocarps were once the dominant forest trees in much of southeast Asia, but the highly prized wood of these species resulted in their decimation. Some dipterocarp species are rare and endemic to certain small areas and are therefore protected. Consequently, poachers attempt to transport logs of these rare species, which resemble common species, once cut. Using DNA barcodes, however, it is possible to differentiate among dipterocarp species and detect logs of rare species that have been illegally harvested (Finkeldey et al., 2010; Dormontt et al., 2015). Various other endangered plants and animals can also be smuggled across borders, and DNA sequences and use of the Tree of Life for comparison can similarly be used to identify species listed in the Convention on International Trade of Endangered Species (CITES) appendices (Lahaye et al., 2008).

Another use of the Tree of Life illustrates the enormous power of "tree knowledge" in a way that would have been hard to imagine or anticipate even a few decades ago. In aquatic ecosystems, assessment of the fauna living there (e.g., fish and amphibians) has typically required netting or electrical shock methods that temporarily stunned organisms so they could be counted and inventoried. Both processes, however, are harmful methods of census for many organisms and perhaps not the best way to assess the population sizes of rare species.

Enter environmental DNA as an important diagnostic tool to probe aquatic ecosystems and determine the fish and amphibian species that are present in a nonintrusive manner. As aquatic organisms swim, they naturally slough off cells into the environment. DNA detection methods are very sensitive—so sensitive that it is possible to sample water from an aquatic habitat and examine the DNA fragments in that sample of water via amplification and DNA sequencing and then match those DNA sequences against data for living aquatic organisms, allowing one to assess which species in the Tree of Life are present in a particular aquatic location. In this way, the presence or absence of rare species can be assessed without disturbing them, and the existence of introduced species can also be determined without harm to the native species (e.g., Valentini et al., 2016; Olds et al., 2016; Evans et al., 2016).

ECOSYSTEM SERVICES

It is clear that healthy ecosystems rich in biodiversity play a major role in our own survival. In other words, the Tree of Life and biodiversity are also fundamental to what are termed "ecosystem services" (Naeem et al., 2009; Costanza et al., 2014). These are broadly defined as the numerous benefits *Homo sapiens* naturally obtains from properly functioning ecosystems. Some of these ecosystem functions seem obvious once we consider them and include things such as clean water, fertile soils, and pollination of plants (including our crops). Healthy ecosystems also reduce flooding, moderate climate, provide clean air, and reduce disease. There are additional recreational and spiritual (mental health) benefits to healthy ecosystems.

Four broad categories of ecosystem services are recognized (Table 5.1): *provisioning*, including providing water and food; *regulating*, such as the control of climate and even disease; *supporting*, such as various nutrient cycles and natural pollination; and *cultural*, which includes recreational opportunities and mental health (Holzman, 2012).

Areas that represent healthy ecosystems are each home to a broad swathe of the Tree of Life. However, during the Anthropocene, rapid changes have occurred in the structure and function of ecosystems, and the pace of ecosystem degradation continues to accelerate (see Chapter 6). As leaves (species) and even entire branches of the Tree of Life are lost from ecosystems, the ability of those ecosystems to function and provide these diverse ecosystem services we humans take for granted is compromised or destroyed (e.g., Palmer et al., 2004; Kremen, 2005).

Ecosystem services have largely been unappreciated and typically are not accounted for in terms of economic impact or cost of the benefits they provide. Such underappreciation of the importance of biodiversity and healthy ecosystems, however, has begun to change over the last few decades. Costanza et al. (2014) first provided an estimate for ecosystem services (based on data for 1997), suggesting that on a worldwide basis they could be valued at $\sim$\$33 trillion annually ($>$\$44 trillion today). Although any such numbers are hard to evaluate and can be controversial, they are a step in the right direction. Without a healthy Tree of Life on a worldwide basis, humanity will suffer greatly—a price tag that is too high

Table 5.1 Summary of ecosystem services

Provisioning services. These species maintain the supply of natural products that are required for human survival. These include food, timber, fuel, various fibers used for clothing, water, soil, medicinal plants, and animals.

Regulatory services. These species facilitate the smooth operation or running of our natural world. As examples, these elements filter pollutants to maintain clean air and water; they also help moderate the climate, work to sequester and store carbon. The recycling of waste and dead organic matter is also included here as are the natural controls of organisms considered agricultural pests and vectors of disease.

Supporting services. These are the services that maintain the provisioning and regulatory services noted above. Include here is the formation of soil, the vital process of photosynthesis by which plants convert sunlight into food by using carbon dioxide and releasing oxygen; these services provide for a healthy habitat. Healthy habitats maintain species diversity as well as genetic diversity; both are the crucial framework of all provisioning and regulatory services noted.

Cultural services. These include the diverse benefits to human well-being that come from contact with nature, including positive psychological and spiritual impacts. These benefits all result from diverse aspects of human culture (e.g., hiking, boating, visits to wild and scenic areas, bird watching, fishing, hunting, gardening). These services have clear health benefits (e.g., stress reduction).

Modified from Holzman, D.C., 2012. Accounting for nature's benefits: the dollar value of ecosystem services. Environ. Health Perspect. 120, A152–A157.

to fathom. Therefore, guarding or protecting the Tree of Life is important so as to ensure a healthy environment for the survival of our own species.

MORAL RESPONSIBILITY AND MENTAL HEALING

In the sections above, we have argued for the importance of knowledge of the Tree of Life, largely from a utilitarian standpoint. However, several authors, including Gorke (2003), have argued that a perspective centered on the benefit to our own species is "not only shortsighted but morally bankrupt" (https://www.kobo.com/us/en/ebook/the-death-of-our-planet-s-species). There is intrinsic value to nature and the many species on our planet. In a way, this relates to what some might refer to as the spiritual benefits of the Tree of Life.

The value of biodiversity extends beyond the dollars and cents of the economic value of medicines, crop relatives, and ecological services of clean air and water (Naeem et al., 2009; Costanza et al., 2014). Nature and biodiversity should be protected for the simple reason of their intrinsic beauty and value; even this aspect of biodiversity provides highly significant benefits to our species. Biodiversity provides the inner peace and tranquility that comes from a walk in the woods, a trip to the mountains, or a day spent on a river or lake. Make no mistake, there is an economic benefit to *H. sapiens* for the tranquility that biodiversity provides. Studies have shown the benefits of biodiversity (and the Tree of Life that connects it) to relieving stress and enhancing the quality of life in humans. Nature is our home—we are not originally creatures of enormous concrete cities, but of woodlands and savanna, and when green spaces are provided, residents of large cities often flock to even these seminatural areas.

Several authors have discussed the physical and emotional impacts on humans that result from the loss of biodiversity (see chapters in Chivian and Bernstein, 2008). Although many recent papers and books now espouse the value of the Tree of Life for mental health, the importance of biodiversity from the standpoint of mental health and spiritual well-being has long been recognized and has a rich history tied to some of the major writers, poets, and philosophers in the United States. Early important figures include Ralph Waldo Emerson, who had a large impact on Henry David Thoreau. Thoreau was deeply influenced by nature and the spiritual healing it afforded. His writings of living in nature in the northeastern United States ("Walden"; Thoreau, 2004 [1854]) represent a classic example of the spiritual value of biodiversity. "I went to the woods because I wished to live deliberately, to front only the essential facts of life, and see if I could not learn what it had to teach, and not, when I came to die, discover that I had not lived. I did not wish to live what was not life …."

In the western United States, John Muir's travels and experiences in nature shaped a conservation movement; he cofounded the Sierra Club and influenced generations of people worldwide (http://vault.sierraclub.org/john_muir_exhibit/life/). Muir, who was greatly influenced by Thoreau, similarly espoused the spiritual value of nature, of biodiversity, of the Tree of Life … "Keep close to Nature's heart … and break clear away, once in a while, and climb a mountain or spend a week in the woods. Wash your spirit clean." (Muir, 1918, first printed in 1890; reprinted in 1918).

DEVELOPING COUNTRIES—THE MOST TO LOSE

The areas that have the most to gain from the Tree of Life are often in the biggest jeopardy of species loss and thus have so much more to lose. The tropics are home to the greatest proportion of terrestrial biodiversity, and yet many countries in these same areas are often jeopardizing their own future for short-term gain. In these same areas, poor people living outside of large cities in rural areas or remaining areas of rainforest actually depend on biodiversity (a functional Tree of Life) for their well-being and very survival (e.g., Byg et al., 2007; Mertz et al., 2007; Peters et al., 1989). The long-term goals of conservation efforts and those of native or indigenous peoples are similar—protect the Tree of Life; but the interactions have sometimes been problematic (Dowie, 2009). Native peoples should not have to choose between survival and conservation. As noted by Dowie (2009) and paraphrased here, when conservation efforts and indigenous peoples work together and acknowledge the interplay and interdependence of biodiversity conservation and cultural survival, they can produce a novel and more effective conservation paradigm—this is a crucial realization.

There are many examples of thoughtful conservation efforts that take this approach of protecting native peoples and their cultural heritage while preserving the Tree of Life; those involving ecotourism are a typical case in point. But other examples get less attention, but may be more effective. For example, we consider here the people in the Amazonian area of Guyana. The arapaima, native to the Amazon basin, is the world's largest scaled freshwater fish—they can reach up to 200 kg (400 lb) in weight. Four or five species may currently exist (one or more may be extinct), but they are not well studied so experts are not really sure how many species there are—another great example of our poor understanding of the Tree of Life. Arapaima have long been hunted for food by indigenous peoples of the Amazon. But, with the incursion of Europeans, the arapaima was harvested in large numbers. As a result of this recent overfishing, they are now considered vulnerable to extinction. The species is an unusual, distinctive lineage in the Tree of Life with no close living relatives. However, these fish are also prized for sportsfishing. Recently, catch and release fly fishing was introduced as a means to protect not only this species but also the land of the native peoples where these fish are found—the income and protection of the land also protects the native peoples' way of life. Sports fisherman will pay large amounts to fish for this species. In addition, members of the village serve as fishing guides and cooks. Fishing has made it far more worthwhile to protect the fish than to hunt and eat them (Purnell, 2018).

The loss of biodiversity in developing countries has many sources—destruction of forest for oil palm (Byg et al., 2007) or for soybeans in Brazil (much of which ultimately feeds pigs and chickens for fast food restaurants worldwide) or cattle ranching (Kirby et al., 2006) or timber resources (Fernside, 2005)—most of which then end up in wealthy countries. These activities provide only short-term gain for the people who make those areas home. The longer-term consequences of that damage to the Tree of Life are significant for the peoples in those areas, but the overall global impact is also enormous.

Solutions to the biodiversity (Tree of Life) crisis in developing countries are varied and highly complex and well beyond the scope of this short book. Obvious solutions (but not necessarily easy to implement) include sustainability, ecotourism, and even local recreational fishing and hunting. Without biodiversity, these pastimes or hobbies are not possible.

The world's wealthiest countries already make a considerable investment in protecting biodiversity in developing countries as well as in training scientists from those countries. But, while developed countries now pledge large amounts to help protect biodiversity in developing countries ($\sim\$10$ billion annually as of 2015), these amounts are low when one considers the estimated amount needed to curb the loss of biodiversity ($\sim\$80-200$ billion) (//india.blogs.nytimes.com/2012/10/23/developing-countries-turn-to-each-other-for-conservation/). While funds to protect biodiversity in developing countries, as well as the important investment in training scientists from those countries to go back to help their home countries do make an enormous difference, ultimately more people in the developing countries have to have ownership—more self-directed and owned initiatives that will promote saving the Tree of Life from within rather than primarily from without.

CONCLUSION

As discussed in Chapter 6, what we don't know can hurt us! There are numerous reasons to be concerned about a sixth mass extinction and the rapid loss of species over a relatively short time frame—the loss of organisms that directly or indirectly hold the key to cure a human disease or improve the human condition. Many organisms have a hidden value as components to ecosystems—often underappreciated. As aptly stated by May (2011), "We are astonishingly ignorant about ... how many [species] we can lose yet still maintain ecosystem services that humanity ultimately depends upon." Ecologists have long evoked a rivet hypothesis (Ehrlich and Ehrlich, 1981) to explain the crucial impact of species loss. Imagine an ecosystem as a large complex airplane, held together with many rivets. As more and more rivets are lost from the airplane (ecosystem), eventually there is a crucial point of malfunction, or collapse—the plane crashes; or in our metaphor, the ecosystem collapses. Furthermore, as species in the ecosystem are lost, the rate of extinction itself increases. As wonderfully stated by E.O. Wilson (2016, p. 14), "As more and more species vanish or drop to near extinction, the rate of extinction of the survivors accelerates." "As extinction mounts, biodiversity reaches a tipping point at which the ecosystem collapses." This is a possibility that should concern all of us today (e.g., Diamond, 2011).

Even if one struggles with these biological concepts or ideas, the fact is that we are the dominant organism on our planet—as Wilson (2016), Gorke (2003), and others have argued, do we not have the moral responsibility to care about the fate of other species—and the Tree of Life—to feel the importance of that connectivity to all life ... a connectivity that our ancient ancestors and indigenous peoples today certainly cherish? In addition, although we often stress the direct economic benefits to humans to protect and preserve the Tree of Life—medicines, crop improvement, ecosystem services (fresh water, clean air)—there is more than that at stake. There is the preservation of the intrinsic value of nature itself—that all species matter and have value (Gorke, 2003).

Numerous authors have made a moral argument for conservation and saving the Tree of Life. Great quotes are found in the many pages written on the topic, for example: "no species is more valuable or meaningful than another except in the minds of humans" (Klinkenborg, 2014). The moral argument for shaping our view of biodiversity and the Tree of Life has been made by such legends as Rachel Carson, Aldo Leopold, and E.O. Wilson, with numerous recent thoughtful contributions on the topic (e.g., Kolbert, 2014).

Is there hope? Without a sustained effort, maybe not (Cafaro, 2015). Wilson (2016) argues for setting aside half the world as wild or natural. With the current estimate of only 17% of the Earth protected, we have a long way to go. Humans appear to show little interest in limiting growth to save our own grandchildren, let alone other species (Dowie, 2009).

But humans also show enormous capacity to work together to solve complex problems and effect change. Every person can make a difference; by working together, the enormous challenges to protecting the Tree of Life can be solved. There is hope.

REFERENCES

Anderson, E., 1967. Plants, Man and Life. University of California Press, Berkeley, Los Angeles, CA; London, p. 256.

Allen, J.M., Germain-Aubrey, C.C., Barve, N., Neubig, K.M., Majure, L.C., Laffan, S.W., et al. Spatial phylogenetics of Florida vascular plants: the effects of calibration and uncertainty on diversity estimates. iScience (in press)

Amborella Genome Project, 2013. The complete nuclear genome of Amborella trichopoda: an evolutionary reference genome for the angiosperms. Science 342 (6165).

Avise, J.C., Nelson, W.S., 1989. Molecular genetic relationships of the extinct dusky seaside sparrow. Science 243, 646−648.

Baker, H.G., 1970. Plants and Civilization. Macmillan, London.

Biswas, S., Fan, W., Li, R., Li, S., Ping, W., Li, S., et al., 2016. The development of DNA based methods for the reliable and efficient identification of Nicotiana tabacum in tobacco and its derived products. Int. J. Anal. Chem. 2016, 4352308 https://doi.org/10.1155/2016/4352308.

Bush, R.M., Bender, C.A., Subbarao, K., Cox, N.J., Fitch, W.M., 1999. Predicting the evolution of human influenza A. Science 286, 1921−1925.

Byg, A., Vormisto, J., Balslev, H., 2007. Influence of diversity and road access on palm extraction at landscape scale in SE Ecuador. Biodivers. Conserv. 16, 631−642.

Cafaro, P., 2015. Recent Books on species extinction. Biol. Conserv. 181, 245−257.

Chivian, E., Bernstein, A., 2008. Sustaining Life: How Human Health Depends on Biodiversity, third ed. Oxford University Press, Oxford, UK, p. 568.

Costanza, R., de Groot, R., Sutton, P., van der Ploeg, S., Anderson, S.J., Kubiszewski, I., et al., 2014. Changes in the global value of ecosystem services. Global Environ. Change Hum. Policy Dimens. 26, 152−158.

Coyle, H.M., Palmbach, T., Juliano, N., Ladd, C., Lee, H.C., 2003. An overview of DNA methods for the identification and individualization of marijuana. Croat. Med. J. 44, 315−321.

Cragg, G.M., Newman, D.J., 2013. Natural products: a continuing source of novel drug leads. Biochim. Biophys. Acta—Gen. Sub. 1830, 3670−3695.

Cui, H., Shi, Y., Ruan, T., Li, X., Teng, Q., Chen, H., et al., 2016. Phylogenetic analysis and pathogenicity of H3 subtype avian influenza viruses isolated from live poultry markets in China. Sci. Rep. 6, 27360. Available from: https://doi.org/10.1038/srep27360.

de Casas, R.R., Soltis, P.S., Mort, M.E., Soltis, D.E., 2016. The influence of habitat on the evolution of plants: a case study across Saxifragales. Ann Bot. 18, 1317−1328.

Diamond, J., 2011. Collapse: How Societies Choose to Fail or Survive. Penguin Books.

Dobzhansky, T., 2013. Nothing in biology makes sense except in the light of evolution. Am. Biol. Teach. 75, 87−91.

Dormontt, E.E., Boner, M., Braun, B., Breulmann, G., Degen, B., Espinoza, E., et al., 2015. Forensic timber identification: it's time to integrate disciplines to combat illegal logging. Biol. Conserv. 191, 790−798.

Dowie, M., 2009. Conservation Refugees: The Hundred-Year Conflict Between Global Conservation and Native Peoples. MIT Press, Boston, MA.

Ehrlich, P.R., Ehrlich, A.H., 1981. Extinction: The Causes and Consequences of the Disappearance of Species. Random House, New York.

Eickmann, M., Becker, S., Klenk, H.D., Doerr, H.W., Stadler, K., Censini, S., et al., 2003. Phylogeny of the SARS coronavirus. Science 302, 1504−1505.

Eilenberg, H., Pnini-Cohen, S., Schuster, S., Movtchan, A., Zilberstein, A., 2006. Isolation and characterization of chitinase genes from pitchers of the carnivorous plant *Nepenthes khasiana*. J. Exp. Bot. 57, 2775−2784.

Evans, N.T., Olds, B.P., Renshaw, M.A., Turner, C.R., Li, Y., Jerde, C.L., et al., 2016. Quantification of mesocosm fish and amphibian species diversity via environmental DNA metabarcoding. Mol. Ecol. Resour. 16, 29−41.

Faith, D.P., 1992. Conservation evaluation and phylogenetic diversity. Biol. Conserv. 61, 1−10.

Fernside, P.M., 2005. Deforestation in Brazilian Amazonia: history, rates, and consequences. Conserv. Biol. 19, 680−688.

Finkeldey, R., Leinemann, L., Gailing, O., 2010. Molecular genetic tools to infer the origin of forest plants and wood. Appl. Microbiol. Biotechnol. 85, 1251−1258.

Fitch, W.M., Bush, R.M., Bender, C.A., Cox, N.J., 1997. Long term trends in the evolution of H (3) HA1 human influenza type A. Proc. Natl. Acad. Sci. U.S.A. 94, 7712−7718.

Forest, F., Grenyer, R., Rouget, M., Davies, T.J., Cowling, R.M., Faith, D.P., et al., 2007. Preserving the evolutionary potential of floras in biodiversity hotspots. Nature 445, 757−760.

Givnish, T.J., 2017. A New World of plants. Science 358, 1535−1536.

Gorke, M., 2003. The Death of Our Planet's Species. Island Press, Washington.

Grifo, F., Rosenthal, J., 1997. Biodiversity and Human Health. Island Press, Washington, DC.

Guan, Y., Zheng, B., He, Y., Liu, X., Zhuang, Z., Cheung, C., et al., 2003. Isolation and characterization of viruses related to the SARS coronavirus from animals in southern China. Science 302, 276−278.

Handa, T., Kita, K., Wongsawad, P., Kurashige, Y., Yukawa, T., 2006. Molecular Phylogeny-Assisted Breeding of Ornamentals. J. Crop Improv. 17, 51−68. Available from: https://doi.org/10.1300/J411v17n01_03.

Harris, C.S., Asim, M., Saleem, A., Haddad, P.S., Arnason, J.T., Bennett, S.A.L., 2012. Characterizing the cytoprotective activity of *Sarracenia purpurea* L., a medicinal plant that inhibits glucotoxicity in PC12 cells. BMC Complement. Altern. Med. 12, 245. Available from: http://www.biomedcentral.com/1472-6882/12/245.

Holzman, D.C., 2012. Accounting for nature's benefits: the dollar value of ecosystem services. Environ. Health Perspect 120 (4), A152−A157.

Hotti, H., Gopalacharyulu, P., Seppanen-Laakso, T., Rischer, H., 2017. Metabolite profiling of the carnivorous pitcher plants *Darlingtonia* and *Sarracenia*. PLoS One 12 (2), e0171078. <https://doi.org/10.1371/journal.pone.0171078>.

International Human Genome Sequencing Consortium, 2004. (Collins, F. S., E. S. Lander, J. Rogers, R. H. Waterston et al.). Finishing the euchromatic sequence of the human genome. Nature 431, 931−945.

Jordan, M.A., Wilson, L., 2004. Microtubules as a target for anticancer drugs. Nat. Rev. Cancer 4, 253−265.

Judd, W.S., Soltis, D.E., Soltis, P.S., Ionta, G., 2007. *Tolmiea diplomenziesii*: a new species from the Pacific Northwest and the diploid sister taxon of the autotetraploid *T. menziesii* (Saxifragaceae). Brittonia 59, 217−225.

Katz, L., Baltz, R.H., 2016. Natural product discovery: past, present, and future. J. Ind. Microbiol. Biotechnol. 43, 155−176.

Kirby, K.R., Laurance, W.F., Albernaz, A.K., Schroth, G., Fearnside, P.M., Bergen, S., et al., 2006. The future of deforestation in the Brazilian Amazon. Futures 38,

432–453. Available from: https://doi.org/10.1016/j.futures.2005.07.011. ISSN 0016-3287.

Klein, N.K., Burns, K.J., Hackett, S.J., Griffiths, C.S., 2004. Molecular phylogenetic relationships among the wood warblers (Parulidae) and historical biogeography in the Caribbean Basin. J. Caribbean Ornithol. 17, 3–17.

Klinkenborg, V. 2014. How to Destroy Species, Including Us. The New York Review of Books. March 20, 2014. www.nybooks.com/articles/2014/03/20/how-to-destroy-species-including-us/.

Kolbert, E., 2014. The Sixth Extinction: An Unnatural History. Bloomsbury Publishing, London, UK, p. 319.

Kuntner, M., Năpăru -Aljančič, M., Li, D., Coddington, J., 2014. Phylogeny predicts future habitat shifts due to climate change. PLoS One 9, e98907. Available from: https://doi.org/10.1371/journal.pone.0098907.

Lahaye, R., Van der Bank, M., Bogarin, D., Warner, J., Pupulin, F., Gigot, G., et al., 2008. DNA barcoding the floras of biodiversity hotspots. Proc. Natl. Acad. Sci. U.S. A. 105, 2923–2928.

Lawing, A.M., Polly, P.D., 2011. Pleistocene climate, phylogeny, and climate envelope models: an integrative approach to better understand species' response to climate change. PLoS One 2011 (6(12)), e28554. Available from: https://doi.org/10.1371/journal.pone.0028554.

Linacre, A., Thorpe, J., 1998. Detection and identification of cannabis by DNA. Forensic Sci. Int. 91, 71–76.

Lu, L.M., Mao, L., Yang, T., Ye, J.F., Liu, B., Li, H.L., et al., 2018. Evolutionary history of the angiosperm flora of China. Nature 554, 234–238.

Magallon, S., Gomez, S., Sánchez Reyes, L.L., Hernández-Hern Andez, T., 2015. A metacalibrated time-tree documents the early rise of flowering plant phylogenetic diversity. New Phytol. 207. Available from: https://doi.org/10.1111/nph.13264.

May, R.M., 2011. Why worry about how many species and their loss? PLoS Biol. 9 (8), e1001130. Available from: https://doi.org/10.1371/journal.pbio.1001130.

McIntosh, M., Cruz, L.J., Hunkapiller, M.W., Gray, W.R., Olivera, B.M., 1982. Isolation and structure of a peptide toxin from the marine snail *Conus magus*. Arch. Biochem. Biophys. 218, 329–334.

Mertz, O., Ravnborg, H.M., Lovei, G.L., Nielsen, I., Konijnendijk, C.C., 2007. Ecosystem services and biodiversity in developing countries. Biodivers. Conserv. 16, 2729–2737.

Miles, D.H., Kokpol, U., Zalkow, L.H., Steindel, S.J., Nabors, J.B., 1974. Tumor inhibitors I: preliminary investigation of antitumor activity of *Sarracenia flava*. J. Pharm. Sci. 63, 613–615.

Mishler, B.D., Knerr, N., González-Orozco, C.E., Thornhill, A.H., Laffan, S.W., Miller, J.T., 2014. Phylogenetic measures of biodiversity and neo-and paleo-endemism in Australian *Acacia*. Nat. Commun. 5, 4473. Available from: https://doi.org/10.1038/ncomms5473.

Mishra, K., Sharma, P., Diwaker, N., Lilly, G., Singh, S.B., 2013. Plant derived antivirals: a potential source of drug development. J. Virol. Antiviral Res. 2, 2. Available from: https://doi.org/10.4172/2324-8955.1000109.

Muhammad, A., Guerrero-Analco, J.A., Martineau, L.C., Musallam, L., Madiraju, P., Nachar, A., et al., 2012. Antidiabetic compounds from *Sarracenia purpurea* used traditionally by the Eeyou Istchee Cree First Nation. J. Nat. Prod. 75, 1284–1288.

Muir, J., 1918. In: Bade, W.F. (Ed.), Steep Trails. Houghton, Mifflin, Boston, MA, p. 390.

Naeem, S., Bunker, D.E., Hector, A., Loreau, M., Perrings, C., 2009. Introduction: the ecological and social implications of changing biodiversity. An overview of a decade of biodiversity and ecosystem functioning research. In: Naeem, S., et al., (Eds.), Biodiversity, Ecosystem Functioning, and Human Wellbeing: An Ecological and Economic Perspective. Oxford University Press, Oxford, UK, pp. 3–13. <https://doi.org/10.1093/acprof:oso/9780199547951.001.0001>.

Noss, R.F., Platt, W.J., Sorrie, B.A., Weakley, A.S., Means, D.B., Costanza, J., et al., 2015. How global biodiversity hotspots may go unrecognized: lessons from the North American Coastal Plain. Diversity Distrib. 21, 236–244.

Olds, B.P., Jerde, C.L., Renshaw, M.A., Li, Y., Evans, N.T., Turner, C.R., et al., 2016. Estimating species richness using environmental DNA. Ecol. Evol. 6, 4214–4226.

Ou, C.-Y., Ciesielski, C.A., Myers, G., Bandea, C.I., Luo, C.-C., Korber, B., et al., 1992. Molecular epidemiology of HIV transmission in a dental practice. Science 256, 1165–1171.

Patwardhan, B., Mashelkar, R.A., 2009. Traditional medicine-inspired approaches to drug discovery: can Ayurveda show the way forward? Drug Discov Today 14, 804–811.

Pennisi, E., 2017. Untangling spider biology. Science 358, 288–291.

Peters, C.M., Gentry, A.H., Mendelsohn, R.O., 1989. Valuation of an Amazonian rainforest. Nature 339, 655–656.

Prie, V., Puillandre, N., Bouchet, P., 2012. Bad taxonomy can kill: molecular reevaluation of Unio mancus Lamarck, 1819 (Bivalvia: Unionidae) and its accepted subspecies. Knowl. Manage. Aquat. Ecosyst 405:18.

Proksch, P., Edrada, R.A., Ebel, R., 2002. Drugs from the seas—current status and microbiological implications. Appl. Microbiol. Biotechnol. 59, 125–134.

Purnell, R., 2018. One guy with a flyrod. Flyfisherman 49 (1), 40–47.

Purvis, A., Gittleman, J.L., Brooks, T.M., 2005. Phylogeny and conservation, Conservation Biology Series, vol. 8. Cambridge, pp. 1–16.

Rahman, A.M., Akter, M., 2015. Taxonomy and traditional medicinal uses of Apocynaceae (Dogbane) family of Rajshahi district, Bangladesh. Res. Rev. J. Bot. Sci. 4, 1–12.

Ruan, B.F., Zhu, H.L., 2012. The chemistry and biology of the Bryostatins: potential PKC inhibitors in clinical development. Curr. Med. Chem. 19, 2652–2664.

Rubio de Casas, R.R., Mort, M.E., Soltis, D.E., 2016. The influence of habitat on the evolution of plants: a case study across Saxifragales. Ann Bot. 18 (7), 1317–1328.

Sala, O.E., Meyerson, L.A., Parmesan, C. (Eds.), 2012. Biodiversity Change and Human Health: From Ecosystem Services to Spread of Disease. Scientific Committee on Problems of the Environment (SCOPE) Series. first ed. Island Press, Washington, DC.

Service, R.F., 2017. Silken promises. Science 358, 293–294.

Shah, V.V., Shah, N.D., Patrekar, P.V., 2013. Medicinal plants from Solanaceae family. Res. J. Pharm. Technol. 6, 143–151.

Singh, R., Sharma, M., Joshi, P., Rawat, D.S., 2008. Clinical status of anti-cancer agents derived from marine sources. Anticancer Agents Med. Chem. 8, 603–617.

Skov, M.J., Beck, J.C., de Kater, A., Shopp, G.M., 2007. Nonclinical safety of ziconotide: an intrathecal analgesic of a new pharmaceutical class. Int. J. Toxicol. 26, 411–421.

Smykal, P., Coyne, C.J., Ambrose, M.J., Maxted, N., Schaefer, H., Blair, M.W., et al., 2015. Legume crops phylogeny and genetic diversity for science and breeding. CRC Crit. Rev. Plant Sci. 34, 43–104.

Soltis, D., et al., 2017. Phylogeny and Evolution of the Angiosperms, Revised and Updated edition. University of Chicago Press.

Soltis, D.E., Albert, V.A., Leebens-Mack, J., Palmer, J., Wing, R., dePamphilis, C., et al., 2008. The Amborella Genome Initiative: A genome for understanding the evolution of angiosperms. Genome Biol. 9, 402.

Staginnus, C., Zoerntlein, S., de Meijer, E., 2014. A PCR marker linked to a THCA synthase polymorphism is a reliable tool to discriminate potentially THC-rich plants of *Cannabis sativa* L. J. Forensic Sci. 59, 919—926.

Stein, B.A., Kutner, L.S., Adams, J.S. (Eds.), 2000. Precious Heritage: The Status of Biodiversity in the United States. Oxford University Press, New York.

Stern, D.B., Nallar, E.C., Rathod, J., Crandall, K.A., 2017. DNA barcoding analysis of seafood accuracy in Washington, D.C. restaurants. Peer J 5, e3234. <https://doi.org/10.7717/peerj.3234>.

Takashi, H., Koichi, K., Pheravut, W., Yuji, K., Tomohisa, Y., 2006. Molecular phylogeny-assisted breeding of ornamentals. J. Crop Improv. 17, 51—68.

Tan, G., Gyllenhaal, C., Soejarto, D.D., 2006. Biodiversity as a source of anticancer drugs. Curr. Drug Targets 7, 265—277.

Thoreau, H.D., 2004. In: Cramer, J.S. (Ed.), Walden, A Fully Annotated Edition. Yale University Press, New Haven, CT, p. 400.

Ulloa, C.U., Acevedo-Rodriguez, P., Beck, S., Belgrano, M.J., Bernal, R., Berry, P.E., et al., 2017. An integrated assessment of the vascular plant species of the Americas. Science 358, 1614—1617.

Valentini, A., Taberlet, P., Miaud, C., Civade, R., Herder, J., Thomsen, P.F., et al., 2016. Next-generation monitoring of aquatic biodiversity using environmental DNA metabarcoding. Mol. Ecol. 25, 929—942.

Vázquez, D.P., Gittleman, J.L., 1998. Biodiversity conservation: does phylogeny matter? Curr. Biol. 8, R379—R381.

Visger, C.J., Germain-Aubrey, C., Soltis, P.S., Soltis, D.E., 2017. Niche divergence in *Tolmiea* (Saxifragaceae): using environmental data to develop a testable hypothesis for a diploid-autotetraploid species pair. Am. J. Bot. 103, 1396—1406.

Warner, K., Timme, W., Lowell, B., Hirschfield, M., 2013. Oceana Study Reveals Seafood Fraud Nationwide. Oceana, Washington, DC.

Warren, W.C., Hillier, L.W., Graves, J.A.M., Birney, E., Ponting, C.P., Grutzner, F., et al., 2008. Genome analysis of the platypus reveals unique signatures of evolution. Nature 453, 175-U171.

Willette, D.A., Simmonds, S.E., Cheng, S.H., Esteves, S., Kane, T.L., Nuetzel, H., et al., 2017. Using DNA barcoding to track seafood mislabeling in Los Angeles restaurants. Conserv. Biol. 31, 1076—1085.

Willis, C.G., Ruhfel, B., Primack, R.B., Miller-Rushing, A.J., Davis, C.C., 2008. Phylogenetic patterns of species loss in Thoreau's woods are driven by climate change. Proc. Natl. Acad. Sci. U.S.A. 105, 17029—17033.

Wilson, E.O., 2016. Half-Earth. Liveright Publishing Corporation, New York.

Zhang, J.T., 2002. New drugs derived from medicinal plants. Therapie 57, 137—150.

CHAPTER 6

Fate of the Tree of Life

The hardest nut to crack, of all the difficult nuts of environmental deterioration, is the real human capacity to forget something not now present that was once of considerable importance to our lives, and the obvious inability to miss something we've never experienced. And so from generation to generation the environment becomes less interesting, less diverse, with smaller unexpected content, and our immediate surroundings become depauperate of animals and plants and exuberant human life. What your parents can hardly remember, you will not miss. What you now take for granted, or what is slowly disappearing, your children, not having known, cannot lament.

Daniel Kozlovsky 1974

GOING, GOING, GONE—DISAPPEARING SPECIES

A major threat to our understanding of the Tree of Life is, of course, extinction, the loss of species. Extinction is a normal part of the history of life on Earth. Fossils of extinct organisms abound throughout the geologic timescale of life on our planet. It is estimated that over the roughly 4-billion-year history of life on Earth, more than 99% of all species that ever lived are extinct (Stearns and Stearns, 2000; Newman, 1997; Novacek, 2014). Although extinction is a normal process, there may be considerable variation among species from the time a species first appeared until the time that given species subsequently goes extinct. Scientists have made estimates of the relatively steady background rate of extinction that has typified Earth's history (e.g., De Vos et al., 2015). Rough estimates are that a species typically exists for perhaps 500,000 to a million years (May et al., 1995; Newman, 1997). Another way to present these estimates is loss of species per year—a standard rate prior to human activity is that for every one million species, one species would be lost per year. Speciation typically outpaces extinction, resulting in a rate of origin of new species that exceeds the loss due to extinction.

However, multiple times during Earth's history, catastrophic events have resulted in the extinction of large numbers of species in a short time frame—these are referred to as "mass extinctions." Five mass

The Great Tree of Life
DOI: https://doi.org/10.1016/B978-0-12-812553-3.00006-0

117

extinctions have been recognized during the history of our planet. Probably the most well-known of these mass extinctions, at least to the public, is the major extinction event that occurred at the end of the Cretaceous period (~66 million years ago), referred to as the Cretaceous—Paleogene (K—Pg) mass extinction. The K—Pg extinction is believed to have resulted from the impact of a massive asteroid (Alvarez et al., 1980; Schulte et al., 2010); it is best known for the extinction of most dinosaurs (one lineage survived, birds), although many other lineages of life also experienced a significant extinction of species (e.g., Fawcett et al., 2009; Longrich et al., 2011, 2012; Rehan et al., 2013; Raup and Jablonski, 1993). But other mass extinctions (Fig. 6.1) had an even greater impact in terms of estimated species loss than did the K—Pg extinction. What is generally considered the worst of the mass extinctions occurred near the end of the Permian (~248 million years ago) and may have resulted in the extinction of 96% of the

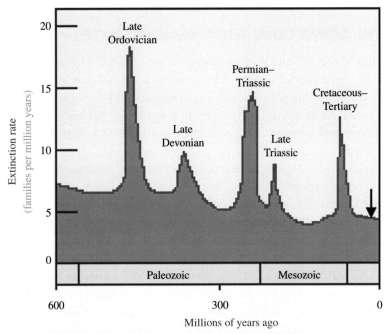

Figure 6.1 There have been five mass extinction events throughout Earth's history. We are now in the early stages of the sixth mass extinction, the Anthropocene (indicated here in a relative sense by the arrow; to be more accurate, the arrow should be almost on top of the right margin). The Cretaceous-Tertiary extinction noted here is the same event referred to in the text as the Cretaceous-Paleogene extinction. *Wikipedia Free Commons.*

species living at that time—as a result, the Permian extinction has sometimes been referred to as "The Great Dying." The Permian extinction is complex, however, and may have actually comprised several phases of extinction events with multiple causes (Benton, 2005).

WELCOME TO THE SIXTH MASS EXTINCTION

Welcome to the early stages of the sixth mass extinction event that is now occurring all across planet Earth (Barnosky et al., 2011; Kolbert, 2014). Many authors have noted that the current pace of extinction of species is very high, much higher than anything seen since earlier mass extinction events (above) in the Earth's history (Barnosky et al., 2011; Pimm et al., 2014; Vignieri, 2014; Ceballos et al., 2015; Payne et al., 2016; Cafaro, 2015; Kolbert, 2014, 2015). The estimates of Ceballos et al. (2015) "reveal an exceptionally rapid loss of biodiversity over the last few centuries, indicating that a sixth mass extinction is already under way." There is no doubt that the sixth mass extinction is largely human-mediated. Perhaps Cafaro (2015) summarizes it best: "The evidence suggests that human beings have been disastrous for other species ever since we evolved into something recognizably ourselves."

Researchers also note that the current high extinction rate of species is not only human-mediated, but also that organisms are being impacted in all environments, with rapid declines in both terrestrial and aquatic habitats, and more recently, marine environments (see below). In fact, the current period is now being referred to as a new geologic period, the Anthropocene. The term is derived from *Anthro*, meaning human, and the ending *cene*, meaning new, as in new period, referring to the human-caused (i.e., anthropogenic) events now shaping global landscapes and impacting biodiversity at what is considered an unprecedented scope and scale. The popular use of the term Anthropocene dates to Crutzen and Stoermer (2000); the term is now widely used in the literature (e.g., Ceballos et al., 2015, 2017; Edgeworth et al., 2015; Lewis and Maslin, 2015; Kolbert, 2014; Wake and Vredenburg, 2008; Barnosky, 2014; Barnosky et al., 2011). The term Anthropocene is now also the subject of popular websites (http://www. smithsonianmag.com/science-nature/what-is-the-anthropocene-and-are-we-in-it-164801414/; http://www.anthropocene.info/short-films.php; www.theguardian.com/environment/2016/aug/29/declare-anthropocene-epoch-experts-urge-geological-congress-human-impact-earth).

Although the term Anthropocene has been only recently accepted and widely used, the term has much older roots. Scientists were concerned about large-scale, human-mediated impact on biodiversity well before the past decade. For example, in 1873, Antonio Stoppani, an Italian geologist, stressed that the human species was having an increased global impact on the Earth—he used the term "anthropozoic era" to refer to the current human-dominated period (see Crutzen, 2002).

There is no official accepted start date for the Anthropocene, although there have been multiple suggestions (Lewis and Maslin, 2015). One proposal is to use as a start date the beginning of the Industrial Revolution in the 1700s and the release of greenhouse gases (Lewis and Maslin, 2015; Edgeworth et al., 2015). But others have suggested that human-mediated changes to soil chemistry may be a better marker of the Anthropocene— human-modified soils are characterized by numerous features, including recurring plowing, addition of fertilizers, various contaminants, leveling, building of embankments, and depletion of organic matter as a result of overgrazing or repeated cultivation (Richter, 2007; Amundson and Jenny, 1991; Zalasiewicz et al., 2010).

Lewis and Maslin (2015) recently evaluated the various proposed dates and diagnostic markers that have been suggested for the start of the Anthropocene. They favor a date in the early 1600s as a start for the Anthropocene for a number of reasons. At that time, major changes occurred on a global basis as a result of the first major collision of human populations and other organisms from the Old World and the New World. The arrival of Europeans in the Americas led to an unprecedented replacement of human populations—nothing vaguely similar has occurred in the past 13,000 years. Native American populations were devastated and largely replaced by people of European, and slightly later, African descent. For the first time, global trade routes linked Europe, the Americas, and China; this linkage had major biological implications. Major food crops, previously localized to specific geographic regions, experienced global patterns of use and distribution, including maize (corn), wheat, potatoes, sugarcane, various legume (bean) crops, and cassava. A similar global movement of domesticated animals also occurred at this time (e.g., chickens, pigs, cows, horses). There was also the first major, global movement of invasive plant and animal species during this period of time. The introduction of the many weedy plants we see commonly in North and South America was initiated at this time. Together, these events all contributed "to a swift ongoing radical reorganization of life on earth without geological precedent" (Lewis and Maslin, 2015), making the early 1600s an excellent choice for the beginning of the Anthropocene.

Lewis and Maslin (2015) proposed 1610 as a very precise start date of the Anthropocene because, in addition to the factors reviewed above, there was a clear dip in atmospheric CO_2 at that time—a distinctive marker that could be used by biologists. The cause of the brief CO_2 dip is an interesting saga on its own. Human global migration spread disease and resulted in a dramatic drop in human populations worldwide, which concomitantly resulted in a drop in agricultural land use and a diminished use of fire by humans, permitting reforestation of many areas. The substantive increase in vegetation during that time and the associated CO_2 uptake resulted in a short-lived global dip in atmospheric CO_2 between the years AD 1570 and AD 1620, a dip that can be readily detected in Antarctic ice cores—a clear, unambiguous marker.

UNDER SIEGE

The Anthropocene has been the age of death for many other species on our planet. The human-mediated demise of other species precedes the estimate (above) of the early 1600s for the start of the Anthropocene. Multiple authors suggest that humans have been driving other species to extinction for many thousands of years (Cafaro, 2015). For example, evidence suggests that our species helped drive large numbers of vertebrate (primarily mammal) species to extinction through hunting near the end of the Pleistocene, $\sim$10,000 years ago (Vignieri, 2014; Gillespie, 2008; Martin, 1967; Barnosky, 2014). During that period, which is termed the great megafauna extinction, perhaps half of all large terrestrial mammals went extinct, including the woolly mammoth and giant ground sloths—and a major contributor seems to have been *Homo sapiens*. That trend of driving other species to extinction has continued and increased in magnitude and breadth. For example, human colonization of tropical Pacific islands in just the past 3000 years led to the loss of $\sim$2000 species and $\sim$8000 island populations of birds (Steadman, 2006). Vertebrate extinctions have been particularly well studied, and over the past 600 years, since before the beginning of the Anthropocene (using the 1610 start date of Lewis and Maslin, 2015), the steady increase in loss of mammals, birds, and other vertebrates is easily seen (Fig. 6.2).

How high are current rates of extinction? The International Union for Conservation of Nature (IUCN) provides an assessment of extinction

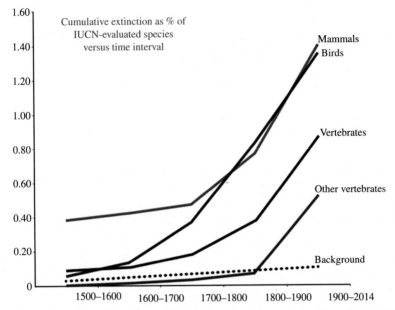

Figure 6.2 Vertebrate extinctions through time, showing accelerating recent extinctions due to human activity. *Modified from Fig. 1 of Ceballos, G., Ehrlich, P.R., Barnosky, A.D., Garcia, A., Pringle, R.M., Palmer, T.M., 2015. Accelerated modern human-induced species losses: entering the sixth mass extinction. Sci. Adv. 1, e1400253.*

rate for species: least concern; near-threatened; three increasing categories of threatened species, "vulnerable," "endangered," and "critically endangered"; and extinct. As summarized by Pimm et al. (2014), as of March 2014, IUCN had assessed 71,576 species that are primarily terrestrial and freshwater: "860 were extinct or extinct in the wild; 21,286 were threatened, with 4286 deemed critically endangered." The percentages of threatened terrestrial species varied, from a low of 13% of bird species, increasing to 31% of reptiles, 37% of fish, and to the highest values of 41% for both amphibians and gymnosperms (nonflowering seed plants; pines and relatives). For flowering plants, 20% of all known species are considered threatened (Brummitt et al., 2008), but considering the huge number of unnamed species (see below), a better estimate for flowering plants is that 30% are threatened (Joppa et al., 2011a). For freshwater species, threat levels range from 23% (freshwater mammals and fishes) to 39% (aquatic reptiles) (Pimm et al., 2014). Some estimates suggest that perhaps half of all species known to science may face extinction by the end of the century (www.theguardian.com/environment/2017/feb/25/half-all-species-extinct-end-century-vatican-conference).

These estimates of extinction risk clearly vary from group to group and suggest that the Anthropocene is impacting and threatening different lineages of life to differing degrees. Perhaps no group of terrestrial organisms has experienced a higher threat to extinction than amphibians—over a third or more of all the ~ 7000 currently living species of amphibians are at risk of extinction. The current amphibian extinction rate may be over 25,000 times the normal background extinction rate (Wake and Vredenburg, 2008; McCallum., 2007). Amphibians have been referred to as the canary in the coal mine as an indicator of the global danger of worldwide extinction, although Kerby et al. (2010) challenge that claim and indicate that while some amphibians may be good environmental indicators, there are other organisms that may be much more sensitive to environmental changes (e.g., plankton).

How do these estimates above compare to what would be considered a typical rate of extinction? Estimates of rates of extinction vary among researchers. However, one widely accepted estimate is that extinction rates are now 1000 times higher than what would be considered a typical background rate, discussed above (De Vos et al., 2015; Pimm et al., 2014). Another way to consider current extinction is that this is roughly equivalent to a loss of 150−200 species every 24 hours (www.theguardian.com/environment/2010/aug/16/nature-economic-security).

More alarming is that investigators agree that extinction will get worse, with future rates of extinction predicted to be 10,000 times higher than normal (De Vos et al., 2015). Again, all of these extinction estimates are much higher than the typical background rate of extinction prior to the Anthropocene.

Other work suggests that because so much of life remains undiscovered and unnamed (as reviewed in Chapter 4, roughly 2.3 million species have been named, but well over 10 million species likely inhabit our planet with some estimates as high as 100 million species), we are actually grossly underestimating the impact of extinction because when extinctions are recorded, they obviously involve only species that have been named—the species that we actually know about. Tedesco et al. (2014) therefore suggest that we are likely underestimating the extinction of undescribed species by 20% or more. For example, if there really are 100 million species on Earth, the current accelerated extinction rate of species loss per year could be as high as between 10,000 and 100,000 species extinctions per year (http://wwf.panda.org/our_work/biodiversity/biodiversity/).

As the dominant species on this planet, we need to ask, in addition to well-documented extinctions that have occurred just within the past 200 years, how many organisms will soon go extinct, which were never known to science, never described, or never named? Thousands of species have already gone extinct as a result of human activity, and many thousands more will go extinct, with many of these species never discovered, never studied, never named—we will never know they existed. Many of those extinct, or soon to be extinct, species may have properties or attributes that could directly or indirectly benefit our own species.

To be clear, making such estimates of current and future extinction is fraught with uncertainty, which scientists, of course, note (Pimm et al., 2014). For example, Pimm et al. (2014) indicate that some earlier estimates (e.g., Barnosky et al., 2011) are probably too high based on the uncertainty involved in several of the measures those workers employed, but largely agree with Tedesco et al. (2014) in which current rates may be 1000 times higher than background. Uncertainty is part of the scientific process. Scientists make it clear that there is variation in projected extinction rates—this variation in hypothesized rates is seen both within and between studies (Pereira et al., 2010). To provide such uncertainty is part of good science. However, good science and noting uncertainty can unfortunately be used by some people to cast doubt on the overall conclusion itself. But, make no mistake, even the low estimates of ongoing extinction rates in the Anthropocene are very high, and projections for extinction in the near future are heading off the charts—these are points of general agreement (Fig. 6.3; Scott, 2008) with a close association of extinction with the skyrocketing growth in the human population worldwide (Fig. 6.3).

The Anthropocene is well underway and to quote an old song, "You ain't seen nothing yet"; these data on extinction rates should be extremely alarming to all humans. Even species not at apparent risk of extinction are now facing sharp declines in population number. Ceballos et al. (2017) point to alarming statistics for terrestrial vertebrates indicating that populations are in decline worldwide, and even species recently considered under low extinction risk or concern are now under threat. In their analysis, 32% of vertebrates are experiencing decline; for the mammals analyzed, all have lost 30% or more of their geographic ranges and greater than 40% have experienced massive population declines. To quote these workers, "beyond global species extinctions Earth is experiencing a huge episode of population declines and extirpations, which will have negative

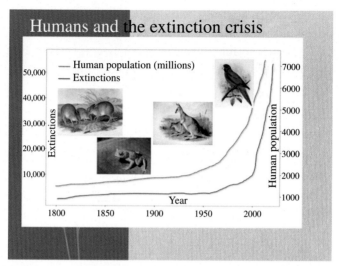

Figure 6.3 Humans and the extinction crisis. Estimated number of species extinctions since 1800 AD (in red) versus human population growth (in green). *Redrawn from Scott, J.M., 2008. Threats to Biological Diversity: Global, Continental, Local. U.S. Geological Survey, Idaho Cooperative Fish and Wildlife, Research Unit, University of Idaho.*

cascading consequences on ecosystem functioning and services vital to sustaining civilization." They refer to this stage of the Anthropocene as "a biological annihilation" to stress the major extent of ongoing species extirpation.

Some people dismiss the extinction crisis stating that "extinction is normal" and add "it gives a chance for other species to evolve and fill in the gaps." There are multiple problems with this human-centric argument. Extinction is indeed "normal." Over the course of millions of years, the formation of new species slightly exceeds the extinction of species. But, make no mistake—the ongoing extinction rate is not "normal," it is human-mediated. An extinction rate 1000 × what is typical (and rising!) is also not "normal," but alarming, representing a new mass extinction period. Previous mass extinctions have dramatically altered the status quo—ecosystems collapsed, the dominant lineages of life disappeared, and ultimately (after millions of years) they were replaced by the rise of other lineages. The rise of the mammals after the extinction of the dinosaurs at the end of the Cretaceous (K—Pg mass extinction) is the best known example to most readers. Such lineage replacement has occurred after each of the previous five mass extinctions. One might ask, what lineages will ultimately replace the now dominant species (*H. sapiens*) after the Anthropocene, should the current trend continue? Furthermore, it also takes many thousands to millions of years for new species to fill the gaps left by extinction—it is not something that occurs quickly.

NO ECOSYSTEM LEFT BEHIND—INTENSIFIED THREATS TO OCEANIC EXTINCTION

While terrestrial biodiversity has been declining rapidly for thousands of years (above), the oceans are now following a similar path (Payne et al., 2016; McCauley et al., 2015). It is inherently more difficult to study and assess both extinction rates and threats to extinction in marine environments than in terrestrial ones (Pimm et al., 2014). The data indicate that threats to the marine environment are relatively recent (Fig. 6.4; McCauley et al., 2015). To date, the human species has been directly responsible for relatively few marine extinctions, but that threat is changing rapidly. The statistics for marine environments are alarming—of 6041 marine species having sufficient data to assess risk, 16% are threatened and 9% near-threatened, most by human overexploitation, habitat loss, and the effects of climate change (The IUCN Red List of Threatened Species, 2014; http://www.iucnredlist.org). Furthermore, marine organisms of large body size seem especially imperiled, which may have particularly problematic downstream effects for ecosystem health and other species (Payne et al., 2016), as can be seen in the saga of the sea otter (below—No Species Is an Island).

Humans have not just changed the terrestrial landscape, they are also dramatically impacting marine wildlife and marine function in every ocean on the planet (McCauley et al., 2015). Current trends indicate that "marine defaunation rates will rapidly intensify as human use of the oceans industrializes" (McCauley et al., 2015). Major threats to marine species are not only human fishing and hunting (of mammals), but in addition habitat degradation is intensifying on a global basis, particularly in response to climate change (McCauley et al., 2015). Perhaps no portion of the marine environment is as important, yet threatened, as coral (see below).

CAUSES OF EXTINCTION

Numerous papers and books have been written on the extent and causes of modern human-mediated extinction and the imperilment of species in

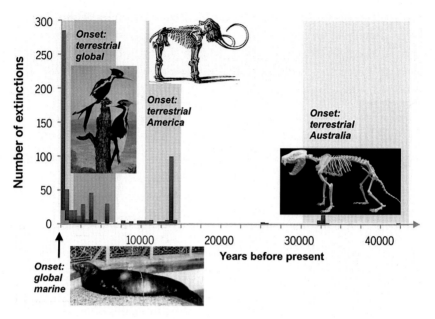

Figure 6.4 No ecosystem left behind. Whereas most human-mediated extinctions through time have been focused on terrestrial organisms, recent human-caused extinctions have now also impacted the oceans. *Modified from Fig. 1 of McCauley, D. J., Pinsky, M.L., Palumbi, S.R., Estes, J.A., Joyce, F.H., Warner, R.R., 2015. Marine defaunation: animal loss in the global ocean. Science 347, 247−253.* Ichthyornis dispar *Marsh (extinct, America recently extinct)* Neomonachus tropicalis *Gray (Caribbean monk seal): https://upload.wikimedia.org/wikipedia/commons/2/21/Cms-newyorkzoologicalsociety1910. jpg;* Campephilus principalis *L. (Ivory-billed woodpecker): https://upload.wikimedia.org/ wikipedia/commons/c/c0/Nature_neighbors%2C_embracing_birds%2C_plants%2C_animals %2C_minerals%2C_in_natural_colors_by_color_photography%2C_containing_articles_by_ Gerald_Alan_Abbott%2C_Dr._Albert_Schneider%2C_William_Kerr_Higley_and_other_ %2814727934526%29.jpg;* Mammuthus primigenius *Blumenbach (extinct, North America, Eurasia): https://ru.wikipedia.org/wiki/Мамонты#/media/File:PSM_V21_D509_ Skeleton_of_a_mammoth.jpg;* Sarcophilus harrisii *Boitard (Tasmanian devil): https://en. wikipedia.org/wiki/Tasmanian_devil#/media/File:Tasmanian_devil_skeleton.jpg.*

the Anthropocene (see review by Cafaro, 2015). It is beyond our scope to review all of the factors that threaten species. We highlight a few of these factors here, habitat loss, climate change, and introduced pathogens. Numerous excellent books and papers have covered the topic of recent extinctions, including works by Wilson and a series of books on individual species extinctions, including works by Greenberg (2014)—the extinction of the passenger pigeon, once the most common bird in

North America, in a few centuries; the recent extinction of the Yangtze River dolphin (Turvey, 2008); the ongoing decimation of our own relatives, the apes (Stanford, 2012), species that may be lost in nature in the next 100 years; as well as the rapid demise of the majestic hemlock (covered below) (Foster et al., 2014).

HABITAT LOSS

Organisms face many threats to their existence, including habitat loss, water and air pollution, climate change, the negative impact of introduced species, and disease. However, the main overriding driver of the loss of species is the tremendous ongoing human population growth taking place at a global scale (Fig. 6.3). The world population of humans for centuries remained fewer than 500,000. The number of humans did not reach 1 billion until about 1800, and then the increase was rapid; in 2017, the human population was roughly 7.5 billion, and 9 billion are projected by 2050. In many regions, human population growth is accompanied by increasing per capita consumption. In sum, human population growth equates to habitat destruction and the decline of many species (Pimm et al., 2014). "How long these trends continue—where and at what rate—will dominate the scenarios of species extinction and challenge efforts to protect biodiversity" (Pimm et al., 2014).

The destruction of natural habitats on a worldwide basis is the major threat to the survival of species (Pimm et al., 2014; International Union for the Conservation of Nature, www.iucn.org). Some figures from the literature are staggering. Forty percent of the terrestrial surface of the Earth is now used for food production (it was estimated to be 7% in 1700) (Vitousek et al., 1997; Sarkar, 2016). In 1997, 50% of the Earth's surface had been modified in some manner for use by *H. sapiens* (Vitousek et al., 1997). Analyses of over 100 countries around the globe showed that human population density predicts with very high accuracy the number of endangered mammals and birds (McKee et al., 2004). As the human population increases in numbers and expands its already enormous global footprint, the result is an increasing demise of species in nature.

Furthermore, our species is also rapidly destroying terrestrial habitats that are considered to contain the most species (Pimm and Raven, 2000). Pimm and Raven (2000) note that perhaps as many as two-thirds of all species occur in the tropics, largely in tropical humid forests. It is estimated that these tropical forests once covered between 14 and 18 million square kilometers—however, only half of that area now remains intact. Loss of rain forests has occurred rapidly in the past several decades, with an approximate loss of 1 million square kilometers every 5–10 years (reviewed in Pimm and Raven, 2000).

But other terrestrial habitats have also seen destruction and species loss, particularly as a result of agricultural land use (Kareiva et al., 2007). A primary habitat for agricultural use is grassland, a major global habitat type also known as prairie, or steppe. Once covering 30% of the Earth's surface, most grasslands have been transformed worldwide for agricultural purposes and grazing. Only 2% of the original grasslands in North America still exist today. Alarmingly, only 8% of existing grasslands on a worldwide basis are protected. Iconic plants and animals reside in prairies, but have seen rapid species depletion and near extinction (in North America, the best example is the American bison).

CLIMATE CHANGE

Scheffers et al. (2016) highlighted the very "broad footprint" of climate change and its immense impact on biodiversity. Climate change has been shown to impact every ecosystem on the planet (Scheffers et al., 2016). Some of the best-known examples of the response of species to climate change involve changes in geographic range. Plants on mountains continue moving upslope to higher altitude, but eventually will run out of real estate (Feeley et al., 2011; Freeman and Freeman, 2014; Brusca et al., 2013). For example, in the past 200 years, plants on the Chimborazo volcano in Ecuador have migrated upslope more than 500 m to higher elevation.

One of the best-documented responses to climate change has been the earlier flowering of plants in temperate regions (Fig. 6.5; Panchen et al., 2012). A critical source of information for tracking the dramatic changes in flowering that are occurring worldwide involves the use of pressed plant specimens (Davis et al., 2015; Willis et al., 2017; Panchen et al., 2012). Botanists have long collected, pressed, and preserved plant specimens by attaching (gluing) them to paper and storing them. There are millions of these pressed plant specimens (also called herbarium specimens; see Fig. 4.7) worldwide, collected over the past several hundred years. These collections can be very helpful in examining when plants flowered in a specific area over long expanses of time. These data can also

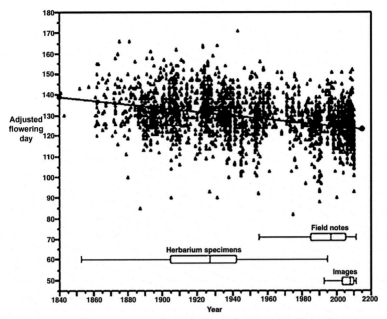

Figure 6.5 Days to flowering time through time, showing that flowering occurs earlier and earlier in response to a changing climate. This example shows a plot of flowering day over time for 28 species in the area of Philadelphia, United States; it is based on a combination of estimates from herbarium specimens (63% of data points; years 1841−2010), field notes (26% of data points; years 1841−2010), and photographic images (11% of data points; years 1977−2010). *From Fig. 1 of Panchen, Z.A., Primack, R.B., Aniśko, T., Lyons, R.E., 2012. Herbarium specimens, photographs, and field observations show Philadelphia area plants are responding to climate change. Am. J. Bot. 99, 751−756. doi:10.3732/ajb.1100198.*

be combined with field notes and photographs. Together, these data indicate that flowering times are becoming earlier and earlier in many areas (Willis et al., 2017; Gaira et al., 2014). For example, plant collection records over a century show temperature-dependent changes in flowering time in the southeastern United States (Park Schwartz, 2015). Similarly, plants in Concord, MA, now flower one week earlier than during the period of 1852-1858, when the famous writer and naturalist Henry David Thoreau made detailed observations of the plants at Walden Pond (Miller-Rushing and Primack, 2008). Miller-Rushing and Primack (2008) determined that these earlier flowering times are correlated with the documented increase in mean temperatures that has occurred in the region one or two months before flowering.

In some cases, however, species that occur together and are closely related respond to climate very differently, indicating the complexity of the responses to climate change. In addition, temporal (timing) shifts due to climate change can have devastating consequences for closely interacting species. For example, climate change can impact nesting cues in birds and result in mismatches with food supplies—an insect prey item that a bird needs to feed its young now emerges earlier than the nesting cycle of the bird (Carey, 2009).

What is also extremely alarming is that many otherwise "healthy" species are also experiencing the local extinction of populations (Ceballos et al., 2017; Wiens, 2016), and recent research indicates that many of these population extinction events are the result of climate change (Wiens, 2016). Furthermore, these population extinction events have already taken place in hundreds of species and across all climate zones and clades of life examined, although the frequency is significantly higher in tropical than in temperate areas. Urban (2015) also found a higher global risk to species in tropical areas due to climate change than in other areas of the globe (Fig. 6.6A; Urban, 2015). As Wiens (2016) notes, this widespread local extinction of populations due to climate change is alarming given that "levels of climate change so far are modest relative to those predicted in the next 100 years." Such climate-related extinctions will certainly become much more prevalent as global warming increases during the next several decades (Wiens, 2016; Urban, 2015). Urban (2015) estimates that extinction risk will greatly increase for one in six species based on climate projections for the next 20 years.

Oftentimes examples of the impact of climate change on only terrestrial ecosystems are given, but an excellent example of the threat of climate change is seen in marine systems and involves coral. Corals are referred to as the rain forests of the oceans because they provide an environment for thousands of other species, many found nowhere else (Stanley, 2003). Most corals are a dual organism consisting of an animal component, a marine invertebrate (Cnidaria), and a photosynthetic unicellular organism, a dinoflagellate (Stanley, 2003). The dinoflagellate provides food to the coral, which enables the coral to grow larger than it would otherwise. The dinoflagellate obtains a home that is a veritable fortress of calcium carbonate deposition. This close symbiotic relationship has been in existence since the Cambrian, 542 million years ago, and became widespread about 100 million years ago (Pratt et al., 2001; Vinn and Motus, 2012; Stanley, 2003). Coral bleaching is a

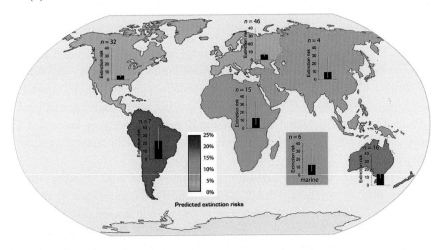

Figure 6.6 (A) The predicted impact (extinction risk; see color scale) of climate change on biodiversity and the Tree of Life varies from region to region on a global basis. Some areas (e.g., tropical areas) are being impacted much more profoundly than other regions of the globe. (B) The large footprint of climate change at different levels of biological organization (e.g., physiology, genetic variation, distribution). *(A) Modified from Fig. 3 of Urban, M.C., 2015. Accelerating extinction risk from climate change. Science 348 (6234), 571–573. (B) Modified from Scheffers, B.R., De Meester, L., Bridge, T.C.L., Hoffmann, A.A., Pandolfi, J.M., Corlett, R.T., Butchart, S.H.M., Pearce-Kelly, P., Kovacs, K.M., Dudgeon, D., Pacifici, M., Rondinini, C., Foden, W.B., Martin, T.G., Mora, C., Bickford, D., Watson, J.E.M., 2016. The broad footprint of climate change from genes to biomes to people. Science 354, 719. Figure obtained courtesy of the authors.*

response to water temperatures that are too warm—the dinoflagellate that is part of the symbiosis ejects from the coral; when the dinoflagellate leaves the coral, the coral loses it beautiful colors and ultimately dies. The lifetime of many coral reefs and the bounty of life they support can now be measured as only a few more decades (see Kolbert, 2014; Hughes et al., 2017), a long-successful marriage destroyed by human-mediated ocean warming (below).

But, as stressed by Scheffers et al. (2016), the impact of climate change goes "beyond well-established shifts in species ranges and changes to phenology and population dynamics to include disruptions that scale from the gene to the ecosystem" (p. 719). Of the 94 ecological processes that they examined, 82% show evidence of impact from climate change (Fig. 6.6B; Scheffers et al., 2016), an alarming statistic. Bottom line—the footprint of climate change on the Tree of Life is immense. For example, climate change is having an impact on the physiology and morphology of species, including shifts in sex ratios in species on land; in aquatic habitats, a decrease in body size has been favored under warmer temperatures. As noted, species are changing in distribution and abundance. Plants from temperate areas are flowering earlier in the year with similar changes in aquatic organisms. Warm-adapted species are experiencing range expansion while cold-adapted species are undergoing dramatic range contraction. As a result, new communities are emerging with tropical species incorporated into temperate areas and boreal species pushed into polar areas.

In addition to the major impact of climate change on biodiversity and the health of the Tree of Life, Scheffers et al. (2016) also stress that many of the effects of climate will directly impact our own species. They note, "The many observed impacts of climate change at different levels of biological organization point toward an increasingly unpredictable future for humans." Impacts are widespread, with an enormous influence economically, and include inconsistent crop yields, decreased productivity in fisheries, decreased fruit yields, and overall threats to food security, changes in the distribution of disease vectors, and the emergence of novel pathogens. Chapter 5 discusses the importance of the Tree of Life to our own species in more detail. As a result of the enormous importance of biodiversity and knowledge of the Tree of Life, rather than simply reacting to climate-induced changes to biodiversity, it is imperative that we are able to predict probable outcomes. In fact, scientists are now working on improved models to better predict the impact of climate change on biodiversity and thus anticipate and attempt to respond (Urban et al., 2016).

"DÉJÀ VU ALL OVER AGAIN": ONE FAMILY'S STORY OF THE DEMISE OF SPECIES

What will it take for people to realize that extinction matters? Have we become immune to reports of species loss and disappearing habitats? Will it be the collapse in the next few decades of coral reef systems and the myriad species they support, including the fisheries that we depend on for food? Perhaps the higher loss of bees and other pollinators because of the direct linkage to crop production? What will it take to notice? Some have argued that one of the best ways to get the point of extinction across is to use a personal touch—connectivity to actual human situations (Cafaro, 2015). Many examples of the demise of iconic species have been well reviewed—these include the ivory bill woodpecker and passenger pigeon. Rather than reviewing these again, we will try a more personal touch—a family story.

To give personal context to the quotation by Daniel Kozlovsky, "What your parents can hardly remember, you will not miss. What you now take for granted, or what is slowly disappearing, your children, not having known, cannot lament," we will take a case history that spans only a few generations in eastern North America. These events have occurred over the course of just three or four human generations. History is quickly repeating itself in this region (as well as in other areas of the world) as one native forest giant after the next is disappearing, typically the result of decimation by an introduced species for which the native tree has no response. In the words of Yogi Berra—"It's like déjà vu all over again." Here are examples experienced by one family—ours.

Until the early 1900s, many forests in eastern North America were dominated in large part by the American chestnut, *Castanea americana*, large, magnificent trees that are in the same family that includes oaks and beech (Faison and Foster, 2014) (Fig. 6.7). The chestnut rapidly disappeared in the early 1900s due to a fungus that was accidentally introduced into the United States from Japan in 1904—by 1940, most trees in eastern North America were gone. The fungus causes an infection (chestnut blight) and killed an estimated 4 billion trees in a few decades. The chestnut was a dominant forest tree. In some areas of the eastern United States, it was estimated that before the chestnut blight, one of every four

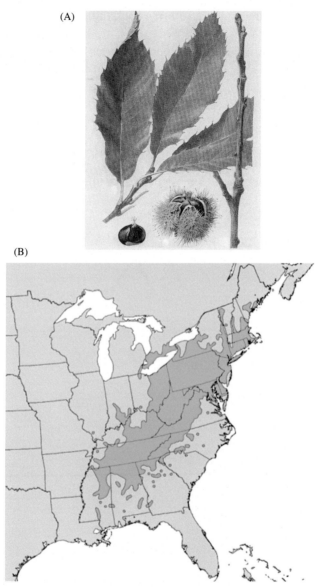

Figure 6.7 (A) American chestnut; *Castanea americana*—plant and (B) American chestnut, historical range in the eastern United States. *Wikipedia Free Commons.*

trees in forested areas was a chestnut (Freinkel, 2007). Our own grandparents saw these chestnut-dominated forests when they were very young, but by the time they were adults, these trees were gone, magnificent trees and forests we have not had the chance to experience (although there are

efforts underway using genetics to bring back the chestnut, https://www.
acf.org/).

Eastern North America next experienced the massive loss of another
major forest tree—the American elm (Fig. 6.8). These were such

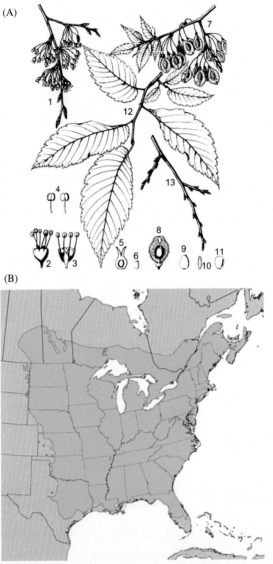

Figure 6.8 (A) American elm; *Ulmus americana*—plant and (B) American elm, historical range in the eastern United States (https://www.forestry.gov.uk/fr/beeh-9u2k3p).
Wikipedia Free Commons.

magnificent trees that they were often planted along city streets—their high arching branches giving a cathedral effect with the street a church-like verdant corridor. Elms, as chestnuts before them, succumbed to a fungus from Asia that was accidentally introduced to both North America and Europe. The disease (Dutch elm disease, so named for Dutch scientists who identified the fungus) is spread by beetles. The disease was first reported from the United States in 1928. Of the estimated 77 million elms that occurred in North America in 1930, over 75% were deceased by 1989 (Hubbes, 1999). Our own parents knew these trees well; although we did not experience the vastness of these forests in our youth, we have witnessed elm trees dying, virtually overnight. Elms in Europe were also devastated. Within 10 years, more than 20 million trees died just in the United Kingdom; in France, over 90% of the elms died. Perhaps 60 million elm trees have been lost to the disease in Europe.

The onslaught to eastern North American woodlands continues as another magnificent giant, the eastern hemlock, *Tsuga canadensis*, is now disappearing from the Appalachian Mountains and New England (Fig. 6.9). The eastern hemlock is a conifer (pine relative) native to eastern North America. These trees can be massive—over 100 ft in height and hundreds of years old. It is the largest conifer in the eastern United States. These immense trees formed expansive forests. We knew these impressive forests well, hiking in the Appalachians and New England as young college students. Now, in many places, only enormous skeletons of these trees remain. This dominant forest tree is victim to the woolly adelgid, a small insect introduced from Asia that sucks the sap from the trees (Foster, 2014). The insect was first introduced in 1924 and then reported in the native range of eastern hemlock in the 1960s. Today, 90% of the range of eastern hemlock has been impacted, and much of the remaining range is also expected to experience insect infection; the stately hemlock will soon be eliminated from most of its native range. Our children and yours will not experience the grandeur of these trees and the forests they once dominated.

But other perhaps lesser known and less obvious species are suffering the same fate as the chestnut, elm, and hemlock. The redbay (*Persea borbonia*), an iconic small tree of the southeastern United States that is in the same genus as the avocado, has been largely exterminated in less than two decades by laurel wilt disease (Hanula et al., 2008; Fraedrich et al., 2008; Koch and Smith, 2008) (Fig. 6.10). The species is being decimated by a

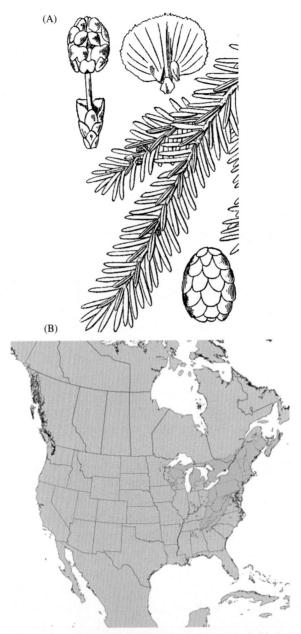

Figure 6.9 (A) Eastern hemlock; *Tsuga canadensis*—plant and (B) Eastern hemlock, historical range in the eastern United States. *Wikipedia Free Commons.*

(A)

(B)   **Distribution of Counties with Laurel Wilt Disease* by year of Initial Detection**

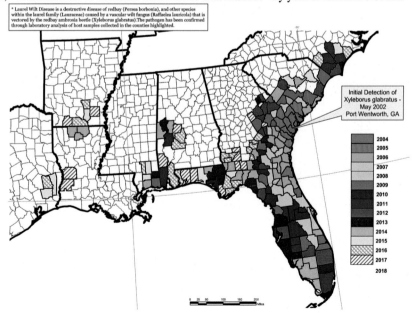

Figure 6.10 (A) Redbay; *Persea borbonia*—plant and (B) Redbay, map showing the spread of the beetle and the fungus it harbors (modified from www.google.com/search?q = spread + of + laurel + wilt + disease&source = lnms&tbm = isch&sa = X&ved = 0ahUKEwjP-fXN0pzdAhUKCawKHZlAAiIQ_AUICigB&biw = 1157&bih = 452#imgrc = p3GDT3fLGR5L6M:). *Wikipedia Free Commons.*

fungus spread by a beetle with which it lives in a symbiotic relationship; the beetle and fungus are native to southeastern Asia and India and were introduced into the southeastern United States—first detected in Georgia in 2002 (Rabaglia, 2008). In Florida, the beetle was first noticed in 2005 (Mayfield and Thomas, 2009). The movement of the beetle through forests is estimated to be up to 34 miles per year (Koch and Smith, 2008). The beetle has spread as far as Mississippi (Riggins et al., 2010), and in its wake, the redbay has disappeared, much as the chestnut was obliterated. This is a tree that was common and that our own children knew; the dead trees still litter forests throughout Florida. Will their children ever have an appreciation of the redbay? Most people in the southeastern United States have not noticed the ongoing decimation and loss of the redbay. But the introduced fungus that is killing redbays also seems to kill other members of the large family Lauraceae (the laurel family) from North America, including the well-known sassafras. The avocado industry in south Florida is also at risk (Fraedrich et al., 2008; Smith et al., 2009), as are hundreds of species of Lauraceae in the New World tropics, should the beetle-fungus duo spread that far.

Just as our generation did not get to experience the grandeur of an American chestnut-dominated forest, trees such as elm, hemlock, and redbay that our generation took for granted will not be here for our children and grandchildren to enjoy or admire. What iconic forest species is next? Species of ash (*Fraxinus*) are now disappearing from our forests due to another invasive, the emerald ash borer (www.nps.gov/articles/ash-tree-update.htm). Over 300 introduced species threaten North America's forests alone (Wilcove, 1999).

NO SPECIES IS AN ISLAND

As many authors have noted, the loss or major demise of a given species will have major impacts on other species, and they too will experience population crash and possible extinction. There are many examples of this cascade or ripple effect—of the loss of one species impacting others and the complexity of species interactions (e.g., Payne et al., 2016; Sandin et al., 2008; Myers et al., 2007; Estes et al., 1989, 1998; Sanders et al., 2018). These ecological cascades can be particularly pronounced when an ecologically dominant species is lost or nearly lost. Of particular concern is the loss of what are termed keystone species—a species that has a very large impact on its surrounding environment relative to its abundance (Paine, 1995).

Several of the examples given in the earlier section attest to the loss of one species impacting many others. The loss of the American chestnut had a dramatic ecological impact. Many animals that were dependent on the chestnut seed for food also experienced dramatic population declines (Davis, 2006; Dalgleish and Swihart, 2012). At least seven species of moth that laid their eggs in leaves of chestnut trees are now considered extinct (Davis, 2006; Opler, 1978; Wilcove, 1999). Species loss can have a measurable economic impact as well. In the case of the chestnut, its economic impact is nicely summarized: "The chestnut tree was possibly the most important natural resource of the Appalachians, providing inhabitants with food, shelter, and a much needed cash income" (Davis, 2006).

Loss of hemlock similarly has had a major ecological impact. These trees form unique forests in eastern North America, with a very moist, low-light forest environment. Many amphibians and birds are associated with these forests, and native brook trout inhabit the streams in these forests; all will be impacted by the loss of hemlock (Siddig et al., 2016; Tingley et al., 2002).

Perhaps a classic example of both the impact of the loss of a keystone species, as well as the overall complexity of species interactions in natural environments, involves sea otters in the Pacific Northwest of North America. In these marine ecosystems, sea otters are the primary predator of sea urchins. Sea urchins in turn consume kelp, large aquatic brown algae that form vast kelp forests and are home to many other species. Without sea otters, sea urchins rapidly destroy massive areas of kelp forests.

However, in the past several decades, sea otter numbers have been collapsing in the Pacific Northwest after an amazing recovery following decades of hunting in the 1800s. Why? The common prey of killer whales had been stellar sea lions and harbor seals. But these species have experienced a collapse ever since the 1970s, probably because of reduction in the abundance of their prey—the fish stocks the sea lions and harbor seals depend on have declined due to a combination of fishing and higher ocean temperatures. The dramatic drop in the numbers of sea lions and harbor seals has, in turn, been accompanied by an apparent shift in predation by killer whales to sea otters, a direct result of the loss of the standard prey for killer whales (Fig. 6.11). The loss of otters subsequently

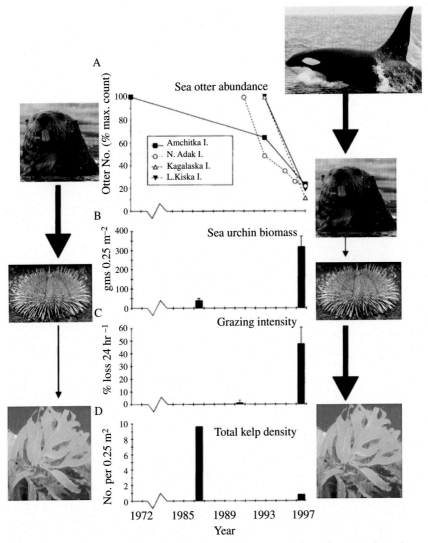

Figure 6.11 The influence on biodiversity and organismal densities based on changes in the number of sea otters. This example of an ecological cascade and the complex interactions among organisms is explained in the text. *Modified from Fig. 1 of Estes, J.A., Tinker, M.T., Williams, T.M., Doak, D.F., 1998. Killer whale predation on sea otters linking oceanic and nearshore ecosystems. Science 282, 473–476.*

had a dramatic impact on kelp forests. Sea kelp forests are now disappearing, as are the organisms that rely on these forests, because without sea otters, the sea urchin population has exploded (Estes et al., 1989, 1998) (Fig. 6.11).

GIVE ME SHELTER—PROTECTED AREAS

We have made enormous progress in designating protected areas. In 1985, less than 4% of all land area was considered to be under some form of legal protection. But, by 2009, that figure had increased to 12.9% (Pimm et al., 2014). And these protected areas, together with other conservation efforts, do make a difference. There are many individual species success stories, usually a charismatic animal, from the American bison to the Miami blue butterfly. There are plant success stories, too, that typically get much less publicity; those plants that do make headlines usually have showy flowers, such as the western prairie fringed orchid (*Platanthera praeclara*; http://www.endangered.org/animal/western-prairie-fringed-orchid/) or rare Hawaiian mint *Stenogyne kanehoana* (www.dvidshub.net/news/260774/army-natural-resources-playing-matchmaker-hawaiis-endangered-plants). But overall total numbers are telling, too, "the rate at which mammals, birds, and amphibians have slid toward extinction over the past four decades would have been 20% higher were it not for conservation efforts" (Hoffmann et al., 2010).

A famous rock song proclaimed, "If I don't get some shelter, I'm going to fade away." So, are we getting it right in our conservation efforts? Humans have been steadily adding to protected areas over the past several decades, but many have argued that current efforts by leaders in some countries to decrease protected natural areas is a major step in the wrong direction. Pimm et al. (2014) stress that while conservation efforts are very important, they are not necessarily representative ecologically—they do not do the best job of protecting biodiversity and the health of the Tree of Life (Pimm et al., 2014). Rodrigues et al. (2004) found that, considering threatened amphibians, birds, mammals, and turtles, that 27%, 20%, 14%, and 10% of the species actually live outside of protected areas. A major global concern is that many species are not found in protected areas at all (Fig. 6.12).

It is also important to remember that most species on our planet remain undiscovered, unnamed—where are they likely located? For plants, there are data suggesting that many of the unnamed species may be in recognized biodiversity hotspots (Joppa et al., 2011b), which seems at first glance to be good news—they are somewhat localized. The following areas combined have 70% of all species that researchers predict remain unnamed and undiscovered: Mexico to Panama (6%), Colombia (6%), Ecuador to Peru (29%), Paraguay and Chile southward (5%),

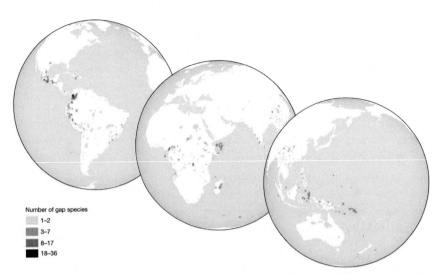

Figure 6.12 Species not covered by protected areas. Using a grid system, researchers (Rodrigues et al., 2004) attempted to assess the areas and density of species not protected by any mechanisms on a global basis. *From Fig. 1 of Rodrigues, A.S.L., Andelman, S.J., Bakarr, M.I., Boitani, L., Brooks, T.M., Cowling, R.M., Fishpool, L.D.C., et al., 2004. Effectiveness of the global protected area network in representing species diversity. Nature 428, 640–643.*

southern Africa (16%), and Australia (8%) (Joppa et al., 2011a). But Joppa et al. (2011b) also note that habitat loss due to human activity is also very extensive in these same important hotspot areas. So, the flip side of the observation that most plant species that remain unnamed are relatively narrowly located, occurring within discrete hotspot areas, is that the risk of extinction of these same species is high. This heightened risk of extinction is especially concerning in that many of the hotspot areas noted are experiencing extensive human-mediated habitat loss (Joppa et al., 2011a).

Some have used a cookie-cutter analogy to discuss species loss. Imagine dough rolled out on a table and random cutting of dough with a cookie-cutter. If a species has a narrow distribution, it obviously has a higher risk of extinction (removal by the cookie-cutter) than does a widespread species. As noted, species are not randomly distributed; we know there are hotspots that contain the highest diversity of species (Chapter 5). Furthermore, the cookie-cutter is not random. Habitat destruction by humans, at least in the tropics, is certainly not random (e.g., Pimm and Raven, 2000; Myers et al., 2000). This, unfortunately, means that a disproportionate amount of dough cutting (habitat destruction and species loss) is actually focused on biodiversity hotspots, meaning that a disproportionate amount of biodiversity is threatened worldwide (Pimm and Raven, 2000; Myers et al., 2000). Myers et al. (2000) note that in 17 tropical forest areas that were designated as biodiversity hotspots, "only 12% of the original primary vegetation remains, compared with about 50% for tropical forests as a whole. Thus, even within those hotspots, the areas richest in endemic plant species have proportionately the least remaining vegetation and the smallest areas currently protected."

The bottom line is well stated by Ceballos et al. (2015)—"Averting a dramatic decay of biodiversity and the subsequent loss of ecosystem services is still possible through intensified conservation efforts, but that window of opportunity is rapidly closing." We can add to that as follows: to have any hope in preserving the Tree of Life, protection of hotspots matters, in that this will result in more "bang for the buck" in terms of species protection. The solution for marine systems seems the same as with terrestrial environments, the designation of more protected areas is crucial (McCauley et al., 2015). These findings give considerable credence to the idea of setting aside more areas that are protected with a focus on biodiversity hotspots. Wilson (2016; see also Hiss, 2014) has argued that half the Earth be set aside in an effort to protect the Tree of Life and sustain a meaningful representation of species on our planet. This is an ambitious goal considering that at present only 15% of the Earth's surface has any type of protection. But as noted above, humans have shown that we can make rapid progress in protecting land (from less than 4% in 1985 to 12.9% in 2009; Pimm et al., 2014)—it can be done. It is a race against time not only for scientists to name a major portion of the many millions of undiscovered and unnamed species on Earth (Costello et al., 2013; Wiens, 2016; Chapter 4), but also for nations to protect the habitats in which these species most likely occur. However, the survival of life as we know it on our planet depends on such bold strategies.

REFERENCES

Alvarez, L.W., Alvarez, W., Asaro, F., Michel, H.V., 1980. Extraterrestrial cause for the cretaceous–tertiary extinction—experimental results and theoretical interpretation. Science 208, 1095–1108.

Amundson, R., Jenny, H., 1991. The place of humans in the state factor theory of ecosystems and their soils. Soil Sci. 151, 99–109.

Barnosky, A.D., 2014. Palaeontological evidence for defining the Anthropocene. Geol. Soc. Spec. Publ. 395, 149–165.

Barnosky, A.D., Matzke, N., Tomiya, S., Wogan, G.O.U., Swartz, B., Quental, T.B., et al., 2011. Has the Earth's sixth mass extinction already arrived? Nature 471, 51–57.

Benton, M.J., 2005. When Life Nearly Died: The Greatest Mass Extinction of All Time. Thames & Hudson, London.

Brummitt, N., Bachman, S.P., Moat, J., 2008. Applications of the IUCN Red List: towards a global barometer for plant diversity. Endangered Species Res. 6, 127–135.

Brusca, R.C., Wiens, J.F., Meyer, W.M., Eble, J., Franklin, K., Overpack, J.T., et al., 2013. Dramatic response to climate change in the Southwest: Robert Whittaker's 1963 Arizona Mountain plant transect revisited. Ecol. Evol. 3, 3307–3319.

Cafaro, P., 2015. Recent Books on species extinction. Biol. Conserv. 181, 245–257.

Carey, C., 2009. The impacts of climate change on the annual cycles of birds. Philos. Trans. R. Soc. Lond. B. Biol. Sci. 364, 3321–3330. Available from: https://doi.org/10.1098/rstb.2009.0182.

Ceballos, G., Ehrlich, P.R., Barnosky, A.D., Garcia, A., Pringle, R.M., Palmer, T.M., 2015. Accelerated modern human-induced species losses: entering the sixth mass extinction. Sci. Adv. 1, e1400253.

Ceballos, G., Ehrlich, P.R., Dirzo, R., 2017. Biological annihilation via the ongoing sixth mass extinction signaled by vertebrate population losses and declines. PNAS 114, E6089–E6096.

Costello, M.J., May, R.M., Stork, N.E., 2013. Can we name earth's species before they go extinct? Science 339, 413–416.

Crutzen, P.J., 2002. Geology of mankind. Nature 415, 23.

Crutzen, P.J., Stoermer, E.F., 2000. The Anthropocene. Global Change Newsl. 41, 17–18.

Dalgleish, H.J., Swihart, R.K., 2012. American chestnut past and future: implications of restoration for resource pulses and consumer populations of eastern US forests. Restor. Ecol. 20, 490–497.

Davis, C.C., Willis, C.G., Connolly, B., Kelly, C., Ellison, A.M., 2015. Herbarium records are reliable sources of phenological change driven by climate and provide novel insights into species' phenological cueing mechanisms. Am. J. Bot. 102, 1599–1609.

Davis, D.E., 2006. Historical significance of American chestnut to Appalachian culture and ecology. In: Steiner, K.C. Carlson, J.F. (Eds.), Proceedings of the Conference on Restoration of American Chestnut of Forest Lands, the North Carolina Arboretum.

De Vos, J.M., Joppa, L.N., Gittleman, J.L., Stephens, P.R., Pimm, S.L., 2015. Estimating the normal background rate of species extinction. Conserv. Biol. 29, 452–462.

Edgeworth, M., Richter, D.D., Waters, C., Haff, P., Neal, C., Price, S.J., 2015. Diachronous beginnings of the Anthropocene: the lower bounding surface of anthropogenic deposits. Anthropocene Rev. 2, 33–58.

Estes, J.A., Duggins, D.O., Rathbun, G.B., 1989. The ecology of extinctions in kelp forest communities. Conserv. Biol. 3, 252–264.

Estes, J.A., Tinker, M.T., Williams, T.M., Doak, D.F., 1998. Killer whale predation on sea otters linking oceanic and nearshore ecosystems. Science 282, 473–476.

Faison, E.K., Foster, D.R., 2014. Did American chestnut really dominate the eastern forest? Arnoldia 72 (2), 18–32.

Fawcett, J.A., Maere, S., van de Peer, Y., 2009. Plants with double genomes might have had a better chance to survive the Cretaceous–Tertiary extinction event. Proc. Natl. Acad. Sci. U.S.A. Available from: https://doi.org/10.1073/PNAS.0900906106.

Feeley, K.J., Silman, M.R., Bush, M.B., Farfan, W., Cabrera, K.G., Malhi, Y., et al., 2011. Upslope migration of Andean trees. J. Biogeogr. 38, 783–791.

Foster, D.R., 2014. Hemlock: a forest giant on the edge. Arnoldia (Jamaica Plain) 71, 12–25.

Foster, D., Baiser, B., Plotkin, A., D'Amato, A., Ellison, A., Orwig, D., et al., 2014. Hemlock: a forest giant on the edge. Yale University Press.

Fraedrich, S.W., Harrington, T.C., Rabaglia, R.J., Ulyshen, M.D., Mayfield III, A.E., Hanula, J.L., et al., 2008. A fungal symbiont of the redbay ambrosia beetle causes a lethal wilt in redbay and other Lauraceae in the southeastern United States. Plant Dis. 92, 215–224.

Freeman, B.G., Freeman, A.M.C., 2014. Rapid upslope shifts in New Guinean birds illustrate strong distributional responses of tropical montane species to global warming. PNAS 111, 4490–4494.

Freinkel, S., 2007. American Chestnut: The Life, Death, and Rebirth of a Perfect Tree. University of California Press, 304 pp.

Gaira, K.S., Rawal, R.S., Rawat, B., Bhatt, I.D., 2014. Impact of climate change on the flowering of *Rhododendron arboreum* in central Himalaya, India. Curr. Sci. 106, 1735–1738.

Gillespie, R., 2008. Updating Martin's global extinction model. Quat. Sci. Rev. 27, 2522–2529.

Greenberg, J., 2014. A Feathered River Across the Sky: The Passenger Pigeon's Flight to Extinction. Bloomsbury Press.

Hanula, J.L., Mayfield III, A.E., Fraedrich, S.W., Rabaglia, R.J., 2008. Biology and host associations of redbay ambrosia beetle (Coleoptera: Curculionidae: Scolytinae), exotic vector of laurel wilt killing redbay trees in the southeastern United States. J. Econ. Entomol. 101, 1276–1286.

Hiss, T., 2014. Can the world really set aside half the planet for wildlife? Smithsonian 45, 66–78.

Hoffmann, M., Hilton-Taylor, C., Angulo, A., Boehm, M., Brooks, T.M., Butchart, S.H. M., et al., 2010. The impact of conservation on the status of the world's vertebrates. Science 330, 1503–1509.

Hubbes, M., 1999. The American elm and Dutch elm disease. For. Chron. 75, 265–273.

Hughes, T.P., Kerry, J.T., Wilson, S.K., 2017. Global warming and recurrent mass bleaching of corals. Nature 543, 373–377.

Joppa, L.N., Roberts, D.L., Myers, N., Pimm, S.L., 2011a. Biodiversity hotspots house most undiscovered plant species. Proc. Natl. Acad. Sci. U.S.A. 108, 13171–13176.

Joppa, L.N., Roberts, D.L., Pimm, S.L., 2011b. How many species of flowering plants are there? Proc. R. Soc. B Biol. Sci. 278, 554–559.

Kareiva, P., Watts, S., McDonald, R., Boucher, T., 2007. Domesticated nature: shaping landscapes and ecosystems for human welfare. Science 316, 1866–1869.

Kerby, J.L., Richards-Hrdlicka, K.L., Storfer, A., Skelly, D.K., 2010. An examination of amphibian sensitivity to environmental contaminants: are amphibians poor canaries? Ecol. Lett. 13, 60–67.

Koch, F.H., Smith, W.D., 2008. Spatio-temporal analysis of *Xyleborus glabratus* (Coleoptera: Circulionidae: Scolytinae) invasion in eastern US forests. Environ. Entomol. 37, 442–452.

Kolbert, E., 2014. The Sixth Extinction: An Unnatural History. Bloomsbury Publishing, London, UK, 319 p.

Kolbert, E., 2015. Field Notes from a Catastrophe: Man, Nature, and Climate Change. Bloomsbury Publishing, USA, 240 pp.

Kozlovsky, D.G., 1974. An Ecological and Evolutionary Ethic. Prentice-Hall, Englewood Cliffs, N.J.

Lewis, S.L., Maslin, M.A., 2015. Defining the Anthropocene. Nature 519, 171−180.

Longrich, N.R., Tokaryk, T., Field, D.J., 2011. Mass extinction of birds at the Cretaceous−Paleogene (K−Pg) boundary. Proc. Natl. Acad. Sci. 108, 15253−15257.

Longrich, N.R., Bhullar, B.-A.S., Gauthier, J.A., 2012. Mass extinction of lizards and snakes at the Cretaceous-Paleogene boundary. Proc. Natl. Acad. Sci. U.S.A. 109, 21396−21401.

Martin, P.S., 1967. Prehistoric overkill. In: Martin, P.S., Wright Jr., H.E. (Eds.), Pleistocene Extinctions: The Search for a Cause. Yale University Press, New Haven, CT, 1967, pp. 75−120.

May, R., Lawton, J., Stork, N., 1995. Assessing Extinction Rates. Oxford University Press.

Mayfield, A.E., Thomas, M.C., 2009. The redbay ambrosia beetle, *Xyleborus glabratus* Eichhoff (Scolytinae: Curculionidae). FDACS-Division of Plant Industry. Available from: http://www.freshfromflorida.com/pi/enpp/ento/x.glabratus.html.

McCallum, M.L., 2007. Amphibian decline or extinction? Current declines dwarf background extinction rate. J. Herpetol. 41, 483−491.

McCauley, D.J., Pinsky, M.L., Palumbi, S.R., Estes, J.A., Joyce, F.H., Warner, R.R., 2015. Marine defaunation: animal loss in the global ocean. Science 347, 247−253.

McKee, J.K., Sciulli, P.W., Fooce, C.D., Waite, T.A., 2004. Forecasting global biodiversity threats associated with human population growth. Biol. Conserv. 115, 161−164.

Miller-Rushing, A.J., Primack, R.B., 2008. Global warming and flowering times in Thoreau's concord: a community perspective. Ecology 89, 332−341.

Myers, R.A., Baum, J.K., Shepherd, T.D., Powers, S.P., Peterson, C.H., 2007. Cascading effects of the loss of apex predatory sharks from a coastal ocean. Science 315, 1846−1850.

Myers, N., Mittermeier, R.A., Mittermeier, C.G., da Fonseca, G.A.B., Kent, J., 2000. Biodiversity hotspots for conservation priorities. Nature 403, 853−858.

Newman, M.E.J., 1997. A model of mass extinction. J. Theor. Biol. 189, 235−252.

Novacek, M.J., November 9, 2014. Prehistory's Brilliant Future. New York Times, p. SR6.

Opler, P.A., 1978, Insects of American chestnut: possible importance and conservation concern In: MacDonald, W.L. (Ed.), Proceedings of the American Chestnut Symposium: Morgantown West Virginia January 4−5, West Virginia University. College of Agriculture and Forestry, Northeastern Forest Experiment Station, Radnor, PA, pp. 83−84.

Paine, R.T., 1995. A conversation on refining the concept of keystone species. Conserv. Biol. 9, 962−964.

Panchen, Z.A., Primack, R.B., Aniśko, T., Lyons, R.E., 2012. Herbarium specimens, photographs, and field observations show Philadelphia area plants are responding to climate change. Am. J. Bot. 99, 751−756. Available from: https://doi.org/10.3732/ajb.1100198.

Park, I.W., Schwartz, M.D., 2015. Long-term herbarium records reveal temperature-dependent changes in flowering phenology in the southeastern USA. Int. J. Biometeorol. 59, 347−355.

Payne, J.L., Bush, A.M., Heim, N.A., Knope, M.L., McCauley, D.J., 2016. Ecological selectivity of the emerging mass extinction in the oceans. Science 353, 1284−1286.

Pereira, H.M., Leadley, P.W., Proenca, V., Alkemade, R., Scharlemann, J.P.W., Fernandez-Manjarres, J.F., et al., 2010. Scenarios for global biodiversity in the 21st century. Science 330, 1496−1501.

Pimm, S.L., Raven, P., 2000. Biodiversity—extinction by numbers. Nature 403, 843−845.

Pimm, S.L., Jenkins, C.N., Abell, R., Brooks, T.M., Gittleman, J.L., Joppa, L.N., et al., 2014. The biodiversity of species and their rates of extinction, distribution, and protection. Science 344, 987.

Pratt, B.R., Spincer, B.R., Wood, R.A., Zhuravlev, A.Y., 2001. Ecology and evolution of Cambrian reefs. In: Zhuravlev, A.Y., Riding, R. (Eds.), The Ecology of the Cambrian Radiation. Columbia University Press, New York, pp. 254–274.

Rabaglia, R., 2008. Xyleborus glabratus. Exotic Forest Pest Information System for North America. http://spfnic.fs.fed.us/exfor/data/pestreports.cfm?pestidval = 148&langdisplay = english (18 April 2011).

Raup, D.M., Jablonski, D., 1993. Geography of end-Cretaceous marine bivalve extinctions. Science. 260, 971–973.

Rehan, S.M., Leys, R., Schwarz, M.P., 2013. First evidence for a massive extinction event affecting bees close to the K-T boundary. PLoS One. 8, e76683.

Richter Jr., D.D., 2007. Humanity's transformation of Earth's soil: Pedology's new frontier. Soil Sci. 172, 957–967.

Riggins, J.J., Hughes, M., Smith, J.A., Mayfield III, A.E., Layton, B., Balbalian, C., et al., 2010. First occurrence of laurel wilt disease caused by *Raffaelea lauricola* on redbay trees in Mississippi. Plant Dis. 94, 634–635.

Rodrigues, A.S.L., Andelman, S.J., Bakarr, M.I., Boitani, L., Brooks, T.M., Cowling, R. M., et al., 2004. Effectiveness of the global protected area network in representing species diversity. Nature 428, 640–643.

Sandin, S.A., Smith, J.E., DeMartini, E.E., Dinsdale, E.A., Donner, S.D., Friedlander, A. M., et al., 2008. Baselines and degradation of coral reefs in the Northern Line Islands. PLOS One 3 (2), 1–11.

Sanders, D., Thébault, E., Kehoe, R., van Veen, F.J.F., 2018. Trophic redundancy reduces vulnerability to extinction cascades. PNAS 115, 2419–2424.

Sarkar, A.N., 2016. Global climate change and confronting the challenges of food security. Productivity 57, 115–122.

Scheffers, B.R., De Meester, L., Bridge, T.C.L., Hoffmann, A.A., Pandolfi, J.M., Corlett, R.T., et al., 2016. The broad footprint of climate change from genes to biomes to people. Science 354, 719.

Schulte, P., Alegret, L., Arenillas, I., Arz, J.A., Barton, P.J., Bown, P.R., et al., 2010. The Chicxulub asteroid impact and mass extinction at the Cretaceous–Paleogene boundary. Science 327, 1214–1218.

Scott, J.M., 2008. Threats to Biological Diversity: Global, Continental, Local. U.S. Geological Survey, Idaho Cooperative Fish and Wildlife, Research Unit, University of Idaho.

Siddig, A.A.H., Ellison, A.M., Mathewson, B.G., 2016. Assessing the impacts of the decline of *Tsuga canadensis* stands on two amphibian species in a New England forest. Ecosphere 7 (11), e01574. Available from: https://doi.org/10.1002/ecs2.1574.

Smith, J.A., Mount, L., Mayfield III, A.E., Bates, C.A., Lamborn, W.A., Fraedrich, S.W., 2009. First report of laurel wilt disease caused by *Raffaelea lauricola* on camphor in Florida and Georgia. Plant Dis. 93, 198.

Stanford, C., 2012. Planet Without Apes. Harvard University Press.

Stanley, G.D., 2003. The evolution of modern corals and their early history. Earth Sci. Rev. 60, 195–225.

Steadman, D.R., 2006. Extinction and Biogeography of Tropical Pacific Birds. The University of Chicago Press, Chicago, IL.

Stearns, B.P., Stearns, S.C., 2000. Watching, From the Edge of Extinction. Yale University Press, 288 pp.

Tedesco, P.A., Bigorne, R., Bogan, A.E., Giam, X., Jezequel, C., Hugueny, B., 2014. Estimating how many undescribed species have gone extinct. Conserv. Biol. 28, 1360–1370.

The IUCN Red List of Threatened Species, 2014. © International Union for Conservation of Nature and Natural Resources. <http://www.iucnredlist.org>.

Tingley, M.W., Orwig, D.A., Field, R., Motzkin, G., 2002. Avian response to removal of a forest dominant: consequences of hemlock woolly adelgid infestations. J. Biogeogr. 29, 1505−1516.

Turvey, S., 2008. Witness to Extinction: How We Failed to Save the Yangtze River Dolphin. Oxford University Press.

Urban, M.C., 2015. Accelerating extinction risk from climate change. Science 348 (6234), 571−573.

Urban, M.C., Bocedi, G., Hendry, A.P., Mihoub, J.-B., Pe'er, G., Singer, A., et al., 2016. Improving the forecast for biodiversity under climate change. Science 353, 1113.

Vignieri, S., 2014. Vanishing fauna. Science 345, 393−396.

Vinn, O., Motus, M.-A., 2012. Diverse early endobiotic coral symbiont assemblage from the Katian (Late Ordovician) of Baltica. Palaeogeogr. Palaeoclimatol. Palaeoecol. 321, 137−141.

Vitousek, P.M., Mooney, H.A., Lubchenco, J., Melillo, J.M., 1997. Human domination of Earth's ecosystems. Science 277, 494−499.

Wake, D.B., Vredenburg, V.T., 2008. Are we in the midst of the sixth mass extinction? A view from the world of Amphibians. Proc. Natl. Acad. Sci. U.S.A. 105, 11466−11473.

Wiens, J.J., 2016. Climate-related local extinctions are already widespread among plant and animal species. PLoS Biol. 14 (12), e2001104. Available from: https://doi.org/10.1371/journal.pbio.2001104.

Wilcove, D.S., 1999. The Condor's Shadow: The Loss and Recovery of Wildlife in America. W.H. Freeman & Company, pp. i−xxii, 1−339.

Willis, C.G., Ellwood, E.R., Primack, R.B., Davis, C.C., Pearson, K.D., Gallinat, A.S., et al., 2017. Old plants, new tricks: phenological research using herbarium specimens. Trends Ecol. Evol. 32, 531−546.

Wilson, E.O., 2016. Half-Earth. Liveright Publishing Corporation, New York.

Zalasiewicz, J., Williams, M., Steffen, W., Crutzen, P., 2010. Response to "The Anthropocene forces us to reconsider adaptationist models of human−environment interactions". Environ. Sci. Technol. 44, 6008.

FURTHER READING

Ehrlich, P.R., Ehrlich, A., 1981. Extinction. The Causes and Consequences of the Disappearance of Species. Random House, New York, pp. i−xiv, 1−305.

Gorke, M., 2003. The Death of Our Planet's Species: A Challenge to Ecology and Ethics. Island Press, Washington, pp. i−xvi, 1−407.

May, R.M., 2011. Why worry about how many species and their loss? PLoS Biol. 9 (8), e1001130. Available from: https://doi.org/10.1371/journal.pbio.1001130.

Peterson Stearns, B., Stearns, S.C., 1999. Watching, from the Edge of Extinction. Yale University Press, pp. i−xv, 1−269.

Pimm, S.L., 2004. A Scientist Audits the Earth, second revised ed. Rutgers University Press, 304 pp.

Pimm, S.L., Russell, G.J., Gittleman, J.L., Brooks, T.M., 1995. The future of biodiversity. Science 269, 347−350.

Royal Botanical Gardens Kew, 2016. State of the world plants. <http://www.kew.org/science/who-we-are-and-what-we-do/strategic-outputs-2020/state-of-the-worlds-plants>.

CHAPTER 7

Teaching the Tree

Human history becomes more and more a race between education and catastrophe.

H. G. Wells 1920

Teach your children well, ... And feed them on your dreams.

G. Nash 1970

song written by G. Nash. First appeared on the album Déjà Vu by Crosby, Stills, Nash & Young released in 1970.

When one uses the phrase "teaching the Tree of Life," it can have several important, interconnected meanings. On the one hand, "teaching the tree" can mean using a tree-based (phylogeny-based) perspective in the classroom (e.g., Ballen and Greene, 2017). This approach is now the accepted and crucial method for teaching students about organismal diversity and relationships. In contrast to focusing on classically used taxonomic group names (e.g., mammals, plants, etc.), using the Tree of Life and "tree-thinking" demonstrates the concept of "family trees" in understanding the organization of biological diversity, that living groups of species are descendants of an ancestor, and that some members of a group are extinct. All of these concepts are illustrated with (phylogenetic) trees. With this teaching approach, students realize that birds are descendants from within reptiles and that birds and dinosaurs share a more recent common ancestor than dinosaurs do with alligators or lizards. This method also reveals that the closest relatives of fungi are animals, not green plants as long taught in the classroom. This critical approach in teaching the Tree of Life and various "tree-thinking" methods to accomplish that goal are reviewed elsewhere (e.g., Ballen and Greene, 2017; Baum and Smith, 2012; Hall, 2011).

A second meaning of "teaching the tree" is to use the Tree of Life as a metaphor for the connectivity of all life—that we are all part of a giant family tree. That knowledge and appreciation will hopefully lead to

The Great Tree of Life
DOI: https://doi.org/10.1016/B978-0-12-812553-3.00007-2
151

increased awareness and interest in protecting the Tree of Life and preserving its many benefits for our own species. In this chapter, the focus lies mainly on the latter meaning of "teaching the tree," but many of the topics are also relevant to the first, more literal teaching of the Tree of Life.

As noted in an earlier chapter, during our short 200,000-year history as a species, we humans long considered ourselves connected to all other species on our planet—representing just a small leaf or branch of the Tree of Life. Many native peoples still maintain that intimate connectivity today. But, as a byproduct of our increasing technological development, humans today often appear disconnected from the other living entities on Earth; a view that perhaps we are a being somehow distinct and separate from the rest of the Tree of Life has become more widespread. This is a perilous mantra. Wilson (2016, p. 1) has summarized well this ongoing decrease in connectivity of our species with the other inhabitants of our planet and the dangerous downstream implications. "What is man? Magnificent in imaginative power and exploratory drive, yet yearning to be more master than steward of a dying planet ... contemptuous towards lower forms of life."

Human connectivity to nature has dropped worldwide in many societies. This is not recent news (Louv, 2005, 2011; see blog review by Choe, 2011). While early humans and many diverse cultures had a firm understanding of the biodiversity in their world, sadly, such connectivity to other species on our planet is often not the case today. These societies knew the plants and other organisms around them and felt interconnected to all of them, and some plants as well as other organisms were even considered sacred. This same connectivity with nature continued in developed countries until very recently, and even through the mid-to-late 20th century, children grew up close to nature, where the experience of a run through the woods or time playing in or near a stream occurred on a regular basis. But as modern societies have changed, these opportunities to connect with nature have diminished. More and more experiences with nature are accomplished via virtual reality.

A prime example of the loss of connectivity with nature is illustrated by the knowledge and outlook of most of our undergraduate botany students at the beginning of a semester. Most of these students see only a blur of green when they see a forest—they rarely see individual entities in those ecosystems. That changes as they begin to recognize individual plant species, and that awakening in awareness is magical to see. We like to tell those same undergrads in the first lecture that basic botany was crucial knowledge for most of the history of our species, and the early humans who failed basic botany failed to become our ancestors. This extreme lack of knowledge regarding nature reveals a dramatic change from the short history of our species through time, and if we continue to fail at teaching the important connection of our species to the other species on our planet, the future of our children and their children is dire indeed.

Teaching the Tree of Life is, therefore, more crucial today than ever. This basic knowledge of the value of the tree and of the connectivity of all life is important for several reasons. First, it is critical to instill in all people the fundamental importance of biodiversity, that each person is part of that Tree of Life—not just an outside viewer or observer, and certainly not at the top of a pyramid of life. Without this sense of connectivity and without a biodiverse world, the quality of the human experience, of human life itself, will certainly decline. It is, therefore, especially important now for people to make this connection to the importance of biodiversity and to take ownership of that connectivity to the Tree of Life. As just one part of the Tree of Life and caretakers of biodiversity, each of us must respond, even in simple ways, to make a difference (www.floridamuseum.ufl.edu/onetree/support/; science.howstuffworks.com/environmental/green-science/save-earth-top-ten.htm). Certainly, this is an important message that we must teach ensuing generations.

The second major reason for teaching the Tree of Life is the point made again and again in this book: that knowledge of relationships matters for our own well-being. From a strictly practical point of view, that information is critical to our own species in terms of new medicines, crop improvement, and overall human health. Imagine the loss of a species to extinction that held the cure to a fatal disease in one of your children, or grandchildren, or best friend—a toolkit of information now lost forever. Given the high (and increasing) extinction rates, such losses are likely happening on a regular basis. In fact, 20%−30% of the plants are now threatened and may be extinct by the end of the century, taking with them their biochemical secrets and potential uses for our own species (Joppa et al., 2011; Brummitt et al., 2015; Royal Botanical Gardens, Kew, 2016; https://stateoftheworldsplants.com/).

With this background in mind, what are the best approaches and methods for "teaching the tree" to students and to the public? There are certainly multiple useful and informative online resources for learning about the Tree of Life, including Encyclopedia of Life, which provides useful compilations of species and knowledge about relationships.

However, there are also challenges in teaching the Tree of Life, such as visualizing large phylogenetic trees that both contain the massive amount of information necessary and are easy to examine and study.

Although hard to believe, for several decades, the only practical means by which scientists could study and display large phylogenetic trees of relationships among species was by printing segments of the tree on consecutive pieces of standard 8-by-10 inch paper and then taping the many pages together in a long linear tree that could be stretched along the floor, sidewalk, or even dropped from large buildings (Fig. 7.1A and B). We can also use these large printed trees to teach the sheer size of the Tree of Life. That is, the actual Tree of Life of all named species, if printed at the same scale (12-point font for species names and normal spacing as in Fig. 7.1A and B) and displayed as a linear (and not a circle) tree, would require four sides of the Empire State Building 14 times to display (Fig. 7.1C).

Although the use of advanced technology has proven helpful, challenges remain in viewing large phylogenetic trees, as they can be simply too big to examine easily by moving the tree image around on a computer screen.

There has been great progress in tree-viewing in recent years, through development of software such as FigTree (Fig. 7.2) (http://tree.bio.ed.ac.uk/software/figtree/) and Dendroscope (http://dendroscope.org/). These tools, however, are not easy to use and are primarily effective tree-viewing tools for experts, not teaching tools for the public. Another effective visualization method is the use of large, integrated walls of video monitors (Fig. 7.3). But once again, this type of resource is not widely available and has severe limitations in reaching the public.

Although big trees are hard to study on a printed page or computer screen, they are nonetheless beautiful and awe-inspiring. They can impress upon the human viewer that our species is just one small branch on the Tree of Life. We can, therefore, provide colorful, one-page circular summary trees of all life that can be wonderful visuals, by placing illustrations of organisms around the tree; these trees can be instructive as well as attractive and striking (Fig. 7.4) and could be made into posters for framing and for display in the classroom or in a museum—or on your wall! While these impressive trees can be viewed as a form of art, they are still hard to use for detailed study and for truly teaching the tree—there are simply too many branches and species.

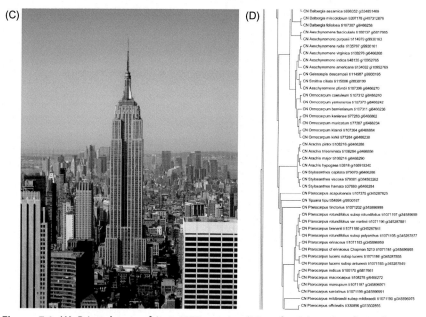

Figure 7.1 (A) Printed tree of just 6000 species (12-pt font) hanging from Century Tower (157 feet tall; 48 meters) at the University of Florida, Gainesville, FL, U.S.A. (B) Same printed tree of only 6000 species stretched along a sidewalk. (C) Empire State Building for scale. The actual Tree of Life of all named species (2.3 million species) if printed at the same scale (12-pt font as in (A) and (B)) would require four sides of the Empire State Building, 14 times to display. See text. (D) One page (a standard 8" X 10" sheet of paper) of the giant tree shown in (A) and (B) showing species names between sample identifiers; e.g., Arachis pintoi.

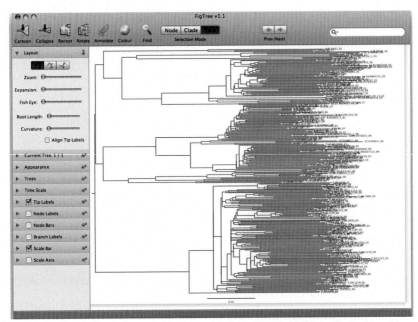

Figure 7.2 Screen view of FigTree (http://beast.community/figtree) showing a phylogenetic tree in a horizontal format. It is possible to expand and zoom in on parts of the tree and search for species. Large trees become harder and harder to navigate, however, as they increase in size (i.e., as the number of species increases).

Figure 7.3 Teaching the Tree of Life (the plant portion of the Tree of Life, modified with images of plants added) using a large network of TV screens available as a teaching tool at the Marston Library, University of Florida. The tree image can be expanded to zoom in on groups of interest.

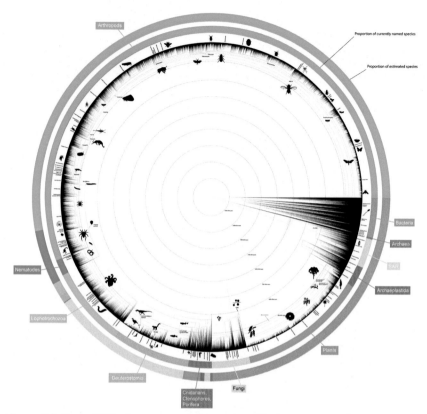

Figure 7.4 Tree of Life shown as a circle tree; modified as a poster showing various groups of organisms around the periphery. *Stephen Smith.*

So, how do we solve the dilemma of giving the public better access to the Tree of Life so that individuals can easily learn about biodiversity and its interconnections? Enter the OneZoom Tree of Life Explorer as an effective tree-viewing tool (http://www.onezoom.org/0). Developed by James Rosindell and colleagues, OneZoom is a very easy-to-use tree viewer for navigating the tree of all 2.3 million named species (published by Hinchliff et al., 2015; see Chapter 3: Building the Tree of Life: A Biodiversity Moonshot) (Fig. 7.5A—OneZoom). OneZoom displays the Tree of Life in an attractive fashion. The Tree of Life looks like an actual tree with a trunk and branches, with species occupying the leaves or tips of the tree. Within each leaf is the species name and a photograph of the organism, if available. Users can zoom into a branch for a closer look (hence the name)—it is a superb teaching tool. The estimated ages of

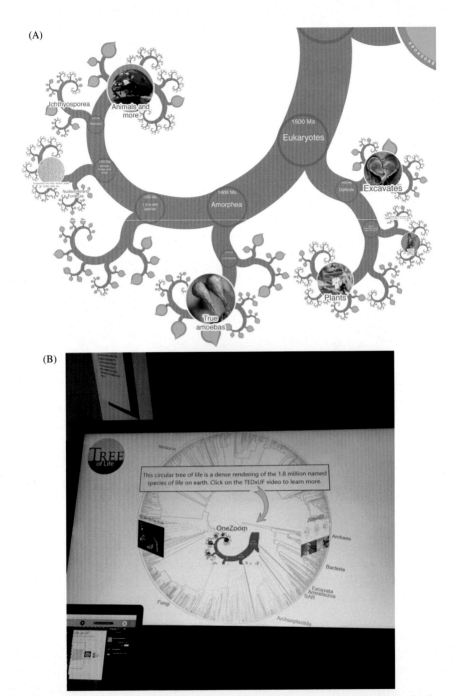

Figure 7.5 (A) Using the OneZoom Explorer to teach and explore the Tree of Life. An image of the OneZoom viewer showing eukaryotes. By clicking on any branch, the person viewing can zoom in on that branch of life and see more detail of relationships, all the way to the species level. (B) OneZoom can be used on a touchtable (as shown here) for tree exploration by students and the public. It is an effective tool in museums and the classroom. In this photograph, a touchtable with the OneZoom Tree of Life has been placed as part of an exhibit in the Harn Museum of Art, University of Florida.

clades are also placed on many branches where such information is available. Because OneZoom can be used on one's own computer or other electronic device, it is highly accessible to the public; plus, OneZoom is easy and fun to use. In addition, the OneZoom visualization tool of the entire Tree of Life can also be placed on a large touchtable (Fig. 7.5B) and used as a teaching tool in the classroom, as well as in natural history museums and other outreach locations.

All tools, however, have their drawbacks, and a limitation of OneZoom is that it also reflects the shortcomings of the Hinchliff et al. (2015) Tree of Life, such as the underrepresentation of the vastness of microbial life (bacteria and archaea). As discussed earlier, this is a result of the fact that scientists today rarely give bacteria and archaea scientific names (see earlier Chapter 4); they are instead issued a molecular accession number. In addition, very little ($<20\%$) of this tree is supported by data beyond the name of the species. The use of OneZoom therefore accurately displays both what we know and do not know about the Tree of Life and can be used to inspire viewers to push the boundaries of current knowledge.

In addition to needing useful tools for teaching the Tree of Life, it is also important to determine effective methods of tapping into public emotion. If people can see clearly that they are intimately connected to all life, perhaps they will want to be more involved in learning more about biodiversity and ultimately taking action to help protect the Tree of Life. The most effective way to teach the tree and make that emotional attachment may be to combine science with the power of music, art, technology, and storytelling in movies and other audiovisual experiences. This type of interdisciplinary approach provides an effective means for stimulating an emotional response that is critical for embracing the importance of the Tree of Life and undertaking the challenge of contributing to its maintenance in a positive way.

A group at the University of Florida and close colleagues has been exploring the ability of two such approaches to have this type of emotional impact on the public: (1) an animated movie and (2) an immense outdoor projection of the Tree of Life on large buildings, similar to a planetarium-like experience in its immensity and beauty (Figs. 7.6–7.8). The goal is to show both products globally to effect change in the public's knowledge of and behavior toward biodiversity.

The movie, *TreeTender*, can be viewed at treetender.org and was produced through a partnership between the Florida Museum of Natural

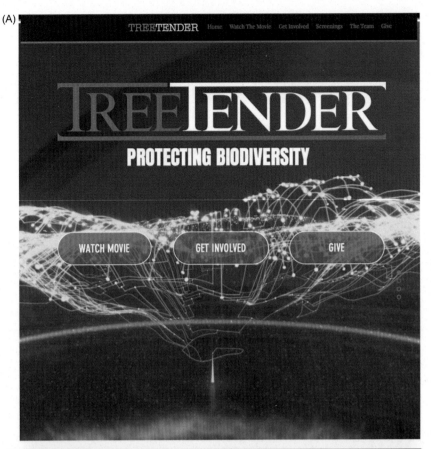

Figure 7.6 (A) TreeTender is an animated movie about the Tree of Life (treetender. org). (B) An audience at the Florida Museum of Natural History watching *TreeTender*. Although the movie can be seen on your computer, it is best appreciated as part of a movie experience.

One Tree—One Planet

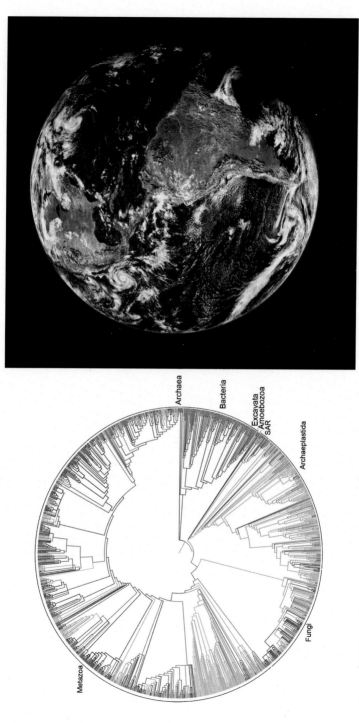

Figure 7.7 One of the images promoting One Tree, One Planet, a projection of the Tree of Life. *Soltis image.*

Archaea

Bacteria

Excavata
Amoebozoa
SAR

Archaeplastida

Fungi

Metazoa

Figure 7.8 (A) A viewer in the audience watching the projection of One Tree, One Planet on a large building. (B) Projection art of the Tree of Life is a moving experience. Members of the audience try to capture a moment of One Tree, One Planet on their cell phones. *Photographs courtesy of the Florida Museum.*

History and the Digital Worlds Institute (both part of the University of Florida). In the style of a Disney movie, *Tree Tender* has two characters (a hero and heroine, if you will) who are immediately engaging to the viewer (Fig. 7.6A and B). The story line not only teaches what the Tree of Life is but why it matters to human health and well-being, through exploration of some basic biological concepts (such as symbiosis) and threats to the Tree of Life (including human-caused extinction). The movie has also been divided into shorter segments that can be used for teaching specific topics (e.g., symbiosis; extinction; ecosystem services). In addition, teaching materials have been prepared for K through 12 classes (all can be found at treetender.org).

The projection of the Tree of Life (referred to as One Tree, One Planet; Fig. 7.7) has a musical score based on variation in DNA sequences in genes that are shared by all life. One Tree, One Planet was developed as a collaboration among renowned projection artist Naziha Mestaoui, who projected images of trees on the Eiffel Tower during the Climate Change Summit in Paris (http://nazihamestaoui.com/one-heart-one-tree-a-virtual-forests-growing-onto-the-citys-famed-monuments/), James Rosindell, the lead developer of OneZoom, and scientists at the Florida Museum of Natural History at the University of Florida. The Tree of Life projection is interactive, and audience members can have their face appear in the place of *Homo sapiens* on a tree that unfolds on a multi-story or larger structure (Fig. 7.8). The projection is truly mesmerizing. With the development of a smart phone app, viewers will gain new opportunities for interacting with the projected Tree of Life as well as with other audience members and past and future viewers around the world, thus enhancing the emotional experience as well as the utility of the projection as a teaching tool.

Will these methods alter people's perspectives and actions? A scientific evaluation team is assessing what works and what does not, providing data to improve these programs. But what we already know is that people leave these experiences somehow changed. *TreeTender* elicits a strong response because of its clear emotional message on the threat and impact of current high extinction rates. Those watching One Tree, One Planet are seen taking frequent photographs of what they see projected (Fig. 7.8B). There is nothing more rewarding than to hear a young child say after watching these pieces: "I am connected to all life." That is a powerful statement and a perspective that is being lost by our species—something we must regain . . . and quickly.

Another way to teach biodiversity is through cell phone apps such as Map of Life (mol.org; Fig. 7.9). This app allows an individual to get a sense of the organisms in the area where he or she is located and will also help identify those species while providing further information. In an age

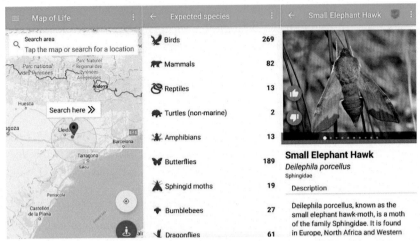

Figure 7.9 Map of Life enables a person to explore aspects of biodiversity using his or her cell phone. It can be used to help identify biodiversity anywhere in the world. It can also be used to record sightings of organisms--these data can then be shared with others. *Wikipedia Free Commons.*

when individuals are often focused on their smart phones, apps such as Map of Life bring nature to people and encourage appreciating and connecting with nature. Map of Life, however, does not show how the species occurring in an area are connected in the great Tree of Life. And this takes us to the idea of the ultimate biodiversity app: one that not only helps the user identify species but also provides information as to where that species fits on the Tree of Life. Perhaps, a link from species identity to tree location in OneZoom could lead to further exploration of biodiversity!

REFERENCES

Ballen, C.J., Greene, H.W., 2017. Walking and talking the Tree of Life: why and how to teach about biodiversity. PLoS Biol. 15 (3), e2001630. Available from: https://doi.org/10.1371/journal.pbio.2001630.

Baum, D.A., Smith, S.D., 2012. Tree Thinking. Freeman, W.H. and Company, New York.

Brummitt, N.A., Bachman, S.P., Griffiths-Lee, J., Lutz, M., Moat, J.F., et al., 2015. Green plants in the red: A baseline global assessment for the IUCN sampled red list index for plants. PLOS ONE 10 (8), e0135152. Available from: https://doi.org/10.1371/journal.pone.0135152.

Choe, R., 2011. Why We Must Reconnect With Nature. Earth Institute, Columbia University, <http://blogs.ei.columbia.edu/2011/05/26/why-we-must-reconnect-with-nature/>.

Hall, B.G., 2011. Phylogenetic Trees Made Easy: A How to Manual. Sinauer Associates.

Hinchliff, C.E., Smith, S.A., Allman, J.F., Burleigh, J.G., Chaudhary, R., Coghill, L.M., et al., 2015. Synthesis of phylogeny and taxonomy into a Comprehensive tree of life. Proc. Natl. Acad. Sci. 112, 12764−12769.

Joppa, L.N., Roberts, D.L., Myers, N., Pimm, S.L., 2011. Biodiversity hotspots house most undiscovered plant species. Proceedings of the National Academy of Sciences of the United States of America 108, 13171−13176.

Louv, R., 2005. Last Child in the Woods: Saving Our Children From Nature-Deficit Disorder. Algonquin Books, Chapel Hill, NC.

Louv, R., 2011. The Nature Principal: Human Restoration and the End of the Nature Deficit Disorder. Algonquin Books, Chapel Hill, NC.

Royal Botanical Gardens, Kew, R. 2016. State of the World's Plants. Available at: https:// stateoftheworldsplants.com/2016/.

Wells, H.G. 1920. Outline of History, 2, ch. 41, pt. 4.

Wilson, E.O., 2016. Half-Earth. Liveright Publishing Corporation, New York.

INDEX

Note: Page numbers followed by "*f*" and "*t*" refer to figures and tables, respectively.

The Great Tree of Life

Douglas E. Soltis
Distinguished Professor, Florida Museum of Natural History, University of Florida

Pamela S. Soltis
Distinguished Professor and Curator, Florida Museum of Natural History, University of Florida

A clear, essential treatise on the Tree of Life, illustrating its historical and cultural connections as well as its many applications

The Great Tree of Life is a concise, approachable treatment that surveys the concept of the Tree of Life, including chapters on its historical use and cultural significance. The "Tree of Life" is a metaphor used to describe the relationships between organisms, both living and extinct and including our own species, as first noted by Charles Darwin. This book expounds on this exciting research and covers how to build an evolutionary tree of relationships. It also demonstrates the enormous utility of the Tree of Life in a variety of applications, including drug discovery, curing disease, crop improvement, conservation biology and ecology, and responding to the challenges of climate change, and discusses the significant ongoing threats to the Tree of Life due to the demands of a growing human population. Lastly, the book stresses the need to find effective ways to "teach" the importance of the Tree of Life to our children, as well as the general public, both to preserve the Tree and to ensure the survival of our own species.

The Great Tree of Life is a key aid to improving our understanding of species relationships and will be of use to researchers, practitioners, academics, and students in biology, ecology, and evolutionary biology and any others interested in understanding the challenges to and opportunities for further conserving and restoring the Earth's biodiversity.

Key Features
- Provides a single reference describing the properties, history, and utility of the Tree of Life
- Introduces phylogenetics and its applications in an approachable manner

Dr. Douglas E. Soltis is a Distinguished Professor in the Florida Museum of Natural History and the Department of Biology at the University of Florida. He has over 400 published papers and eight books. He was president of the Botanical Society of America (1999-2000) and has received the Centennial Award and the Distinguished Fellow Award from the Botanical Society of America, the Asa Gray Award from the American Society of Plant Taxonomists, the Stebbins Medal from the International Association for Plant Taxonomy, and the Darwin-Wallace Medal from the Linnean Society of London.

Dr. Pamela S. Soltis is a Distinguished Professor and Curator in the Florida Museum of Natural History at the University of Florida. She has over 400 published papers and eight books. She served as president of the Society of Systematic Biologists (2004-07), the Botanical Society of America (2006-09), is currently the President-Elect of the American Society of Plant Taxonomists, and has received the Centennial Award and the Distinguished Fellow Award from the Botanical Society of America, the Asa Gray Award from the American Society of Plant Taxonomists, the Stebbins Medal from the International Association for Plant Taxonomy, and the Darwin-Wallace Medal from the Linnean Society of London.

ISBN 978-0-12-812553-3

ACADEMIC PRESS
An imprint of Elsevier
elsevier.com/books-and-journals

9 780128 125533